SpringerBriefs in Environmental Science

T0236454

SpringerBriefs in Environmental Science present concise summaries of cutting-edge research and practical applications across a wide spectrum of environmental fields, with fast turnaround time to publication. Featuring compact volumes of 50 to 125 pages, the series covers a range of content from professional to academic. Monographs of new material are considered for the SpringerBriefs in Environmental Science series.

Typical topics might include: a timely report of state-of-the-art analytical techniques, a bridge between new research results, as published in journal articles and a contextual literature review, a snapshot of a hot or emerging topic, an in-depth case study or technical example, a presentation of core concepts that students must understand in order to make independent contributions, best practices or protocols to be followed, a series of short case studies/debates highlighting a specific angle.

SpringerBriefs in Environmental Science allow authors to present their ideas and readers to absorb them with minimal time investment. Both solicited and unsolicited manuscripts are considered for publication.

More information about this series at http://www.springer.com/series/8868

Pratima Bajpai

Biological Odour Treatment

Pratima Bajpai
C-103 Thapar Centre for Industrial R&D
Patiala
India

ISSN 2191-5547   ISSN 2191-5555 (electronic)
ISBN 978-3-319-07538-9   ISBN 978-3-319-07539-6 (eBook)
DOI 10.1007/978-3-319-07539-6
Springer Cham Heidelberg New York Dordrecht London

Library of Congress Control Number: 2014944807

Printed on acid-free paper

Springer is part of Springer Science+Business Media (www.springer.com)

Preface

Odour is a serious complaint associated with waste air emissions that creates nuisance. Its treatment process ranges from physical and chemical to biological means. A biological treatment system has several advantages over the physical and chemical technologies in being ecofriendly, more efficient with low operational cost and characterized by high flow rates of waste gas with low concentration of contaminants. This book, *Biological Odour Treatment* provides an updated and detailed overview on biological odour treatment and provides comprehensive and in-depth treatment of new technologies. This book will be a valuable reference tool for graduate students, scientists, industrial consultants, biotechnologists, microbiologists and chemical, biochemical, environmental and civil engineers who are interested in environmental sciences, and particularly, in innovative biological technologies for treatment and control of odour and air pollution. I hope that students, teachers, scientists and engineers, will find the descriptive and practical contents of this book interesting and helpful.

Patiala, India   Pratima Bajpai
April, 2014

Contents

Chapter 1
General Introduction

1.1 General Introduction

Odour is certainly the most complex of all the air pollution problems. Odour pollution contributes to photochemical smog formation and particulate secondary contaminant emissions. Therefore it is a threat to human health and welfare and air quality. Odour affects human beings in many ways. Strong, offensive smells, if they are frequent and or persistent, interfere with the enjoyment of life. Foul odour may not cause direct damage to health but toxic stimulants of odour may cause respiratory problems. Very strong odours result in nasal irritation and activation of symptoms in individuals with breathing problems or asthma; they can even prove fatal if people are exposed above a certain limit. Eye irritation has been also reported. Secondary effects may be nausea, fatigue, insomnia, headache and dizziness. Loss of property value near odour-causing industries and odourous environments is partly a result of offensive odour. Growing public awareness of the health and environmental impacts of odour, combined with the implementation of many titles of the US 1990 Clean Air Act Amendments and similar regulations in Europe, have been forcing many industrial and agricultural processes, transport functions, energy production and effluent treatment systems to meet the emission standards laid down in guidelines (Soccol et al. 2003).

The odourous emissions generated from pulp and paper industry has been the cause of nuisance since the inception of the industry (Chan 2006; Burgess et al. 2001). The distinctive odour of sulphur is characteristic of many industrial processes, including the kraft pulp mill process used in the manufacture of paper (Anderson 1970; Andersson et al. 1973; Springer and Courtney 1993). Refineries, sewage treatment plants, gas wells, coke manufacturing plants, chemical manufacturing and leather making also may release sulphur compounds (Nanda et al. 2012; Rappert and Muller 2005). This odour is mostly caused by hydrogen sulphide and to a lesser degree by several other reduced sulphur compounds collectively referred to as 'total reduced sulphur' (TRS) compounds (US EPA 1979). The human nose is particularly sensitive to TRS compounds and is capable of detecting them at concentrations as small as 1 ppb of air.

P. Bajpai, *Biological Odour Treatment*, SpringerBriefs in Environmental Science, DOI 10.1007/978-3-319-07539-6_1, © The Author(s) 2014

Table 1.1 Primary total reduced sulphur (TRS) chemical compounds

Hydrogen Sulphide (H_2S)

It is a colourless, flammable gas with an offensive odour similar to rotten eggs. Hydrogen sulphide emissions originate from the breakdown of sodium sulphide, a component of the kraft cooking liquor. It is a feebly acidic gas which partially ionizes in aqueous solution. The ionization proceeds in two stages with the formation of hydrosulphide and with increasing pH, sulphide ions. Hydrogen sulphide and methyl mercaptan are responsible for the characteristic odour of pulp and paper mills. Hydrogen sulphide generally represents the largest gaseous emissions from the kraft process.

Methyl mercaptan (CH_4S)

It is a gas at normal temperature and is extremely foul smelling (described as decayed cabbage) even at very low concentrations. It is formed during kraft cooking by the reaction of hydrosulphide ions and the methoxy lignin compounds of the wood. Methyl mercaptan dissociates in an aqueous solution to methyl mercaptide ions and this dissociation is completed above a pH of 12.0. CH_3SH is also present in low concentrations as a dissolved gas in the black liquor. As the pH decreases, methyl mercaptan gas is evolved from the black liquor. Methanethiol is another name for this compound. One use of it is as an odourant to make natural gas detectable.

Dimethylsulphide ($(CH_3)_2S$)

It is also known as methylthiomethane and methylsulphide. This is a nonacidic liquid. Dimethyl disulphide is primarily formed through the reaction of methyl mercaptan ion with the lignin component of the wood. It does not dissociate as hydrogen sulphide and methyl mercaptide do. Dimethyl sulphide may also be formed by the disproportion of methyl mercaptan. At normal liquor temperature, it is highly volatile.

Dimethyldisulphide ($(CH_3)_2S_2$)

It is formed by the oxidation of methyl mercaptan throughout the recovery system. It is also known as methyldisulphide and methyldithiomethane. It is also a nonacidic liquid.

The two major components in wood are cellulose and lignin. Lignins are organic compounds that bind the wood cellulose fibres together. In order to make cellulose usable for paper manufacture, the lignins must be separated from the cellulose: a process known as pulping. Kraft pulping is the most commonly used paper pulping process (Smook 1992; Biermann 1996; Gullichsen 2000). Wood chips are cooked in a digester under pressure in a solution of sodium sulphide and sodium hydroxide known as white cooking liquor to separate the lignin and cellulose. The pulp is then filtered, washed, bleached, pressed and dried into paper. The sodium sulphide in the white cooking liquor is the source of the sulphur in the TRS compounds (Bajpai 2008). Although hydrogen sulphide is generally the main TRS compound emitted from kraft pulping, several other compounds are formed during digestion as the sodium sulphide reacts with the lignin in the wood and with process gases. One of these compounds is methyl mercaptan, the highly odiferous substance that natural gas companies purposely add to odourless natural gas to facilitate the detection of gas leaks. The other compounds are dimethyl sulphide and dimethyl disulphide (Table 1.1) (Smet et al. 1998). All these four compounds have low odour thresholds. These are released from many points within a mill, but the main sources are the digester/blow tank systems and the direct contact evaporator. Although most TRS compounds are released from the digester, other parts of the process, such as older design evaporators (in which spent cooking liquor is concentrated for reuse), recovery furnaces and pulp washing and processing can also be sources of TRS

Table 1.2 The major sources of TRS emissions

Digester	Methyl mercaptan, methanol
Black liquor storage tank	Hydrogen sulphide, methyl mercaptan, dimethyl sulphide
Evaporator	Hydrogen sulphide, dimethyl disulphide, dimethyl sulphide, methanol
Recovery boiler	Hydrogen sulphide methyl mercaptan, dimethyl sulphide
Smelt dissolving tank	Hydrogen sulphide, methyl mercaptan
Lime kiln	Methyl mercaptan, sulphur dioxide

Table 1.3 Generation of TRS in the pulp and paper industry

Source	Methyl mercaptan (ppm)	Dimethyl sulphide (ppm)	Methanol (ppm)
Digester relief	500–3000	100–8000	~1000
Digester blow	1000–10,000	2000–20,000	2000–8000
Evaporator vent	2000–45,000	1000–50,00	2000–25,000

emissions. Temporary process upsets or even weather changes can sometimes cause unusually strong but transient TRS odours in the vicinity of these plants.

The major source of TRS emissions include digester blow and relief gases, multiple effect evaporator vents and condensates, recovery furnaces with direct-contact evaporators, smelt dissolving tank and slacker vents, brown-stock washers, seal tank vents and lime kiln exit vents, as shown in Table 1.2. Table 1.3 shows the generation of TRS in pulp and paper industry.

1.2 Health Effects of TRS Compounds

Atmospheric pollution affects the life of millions of peoples in all parts of the world, particularly those who are living in large industrialized cities with unpleasant odours, fumes, dust and corrosive gases, which are harmful to human health, crops and property. A wide variation in the susceptibility of individuals to the effects of air pollution has been reported. Healthy adults are able to endure relatively high concentrations of harmful substances without suffering harm, while the old, young and sick are relatively much more sensitive. The research on TRS has particularly focussed on hydrogen sulphide, which is the most toxic component of the mixture and also the greatest proportionally (Ontario MOE 2007; AMEC 2004; BC MOE 2009). The lower detectable limits for hydrogen sulphide, methyl mercaptan, dimethyl sulphide and dimethyl disulphide are shown in Table 1.4. Hydrogen sulphide has a characteristic "rotten egg" smell. It can be sensed at low levels—0.001–0.13 ppm (CCOHS 2012). Hydrogen sulphide is easily absorbed through the lungs (US EPA 2003). Hessel et al. (1997) report that hydrogen sulphide acts by stopping cellular uptake of oxygen by inhibiting cytochrome oxidase. Within 2–15 min of exposure at levels above 100 ppm , eye irritation and inflammation is observed (CCOHS 2012). Although eye irritation has also been observed at lower concentrations (10 ppm), it is not clear if these effects are due to

Table 1.4 Odour threshold concentration of TRS compounds. (Based on Springer and Courtney 1993)

Reduced sulfur compound	Odour threshold concentration (ppb)
Hydrogen sulfide (H_2S)	8–20
Methyl mercaptan (CH_3SH)	2.4
Dimethyl sulfide (CH_3SCH_3)	1.2
Dimethyl disulfide (CH_3SSCH_3)	15.5

hydrogen sulphide alone, or in combination with exposure to other gases. Exposure to higher levels—100 pm between 2–15 min or continued exposure—may result in olfactory nerve fatigue, making odour itself a poor indicator of the presence of hydrogen sulphide (ATSDR 2011). Irritation of the nose, throat and lungs may result at levels of or above 100 ppm. Olfactory pulmonary oedema has been reported at levels above 250 ppm (CCOHS 2012). Beyond 500 ppm, exposure can cause loss of consciousness, also known as 'knockdown' (ATSDR 2006; Slaughter et al. 2003). At higher levels (500–1000 ppm), tissue hypoxia, cardiovascular effects, central nervous system depression and respiratory arrest can occur, which can result in death (CCOHS 2012; ATSDR 2006). Several long-term, persistent health effects have been reported by individuals who have experienced acute exposure to hydrogen sulphide resulting in 'knockdown' in occupational settings. Typical symptoms include neurological effects such as headaches, impaired memory, problems with focusing, respiratory effects (wheeze, shortness of breath) and ocular dysfunction (corneal abrasions) (Hessel et al. 1997; ATSDR 2006). Damage to brain structures, including to the basal ganglia and cortex, have been reported during follow-up of individuals occupationally exposed to acute levels of hydrogen sulphide (Hessel et al. 1997). Occupational studies have found evidence for adverse health effects, including bronchial hyperresponsiveness and mood disorders.

Long-term health effects in communities exposed to short-term periods of high or long-term periods of low industrial-based emissions have been reported by several researchers (Legator 2001; Partti-Pellinen 1996; Campagna et al. 2004; Haahtela 1992; Marttila 1995; Inserra et al. 2004). Some evidence of respiratory and central nervous system-related effects in residents exposed to ambient TRS and/or hydrogen sulphide has been found. Symptoms include cough, eye and nasal irritation, breathlessness and nausea. These studies have many shortcomings, which make it difficult to judge the level of health risk posed to community members exposed to ambient TRS and/or hydrogen sulphide. Because of the potential chronic nature of exposure in communities, it is difficult to separate acute and chronic exposure-related effects.

Methyl mercaptan affects the central nervous system. It causes paralysis of the respiratory centre and has been found to cause convulsions and narcosis at high concentration. At lower concentrations, it causes pulmonary oedema (Hessel et al. 1997). The toxic effect of methyl mercaptan can be easily understood by studying a case highlighted by Shults et al. (1970), in which a man emptying gas cylinders of methyl mercaptan was overexposed. He was found comatose at the worksite and hospitalized. He developed acute haemolytic anaemia and methemoglobinaemia and remained in a deep coma until dying 28 days after the accident.

Conventional air pollution control technologies can treat a wide variety of pollutants at higher concentrations; however, for treating waste air with low pollutant concentrations these approaches become economically prohibitive. Biological methods for the removal of odours and volatile organic compounds from waste gases are cost-effective technologies when low concentrations are to be dealt with (Ottengraph and Van Denoever 1983; Ottengraf 1986, 1987; Ottengraf et al. 1986; Thorsvold 2011; Brauer 1986; Cloirec et al. 2001; Deshusses 1997; Hort et al. 2009; Chung 2007; Anit and Artuz 2000; Devinny et al. 1999; Farmer 1994; Finn and Spencer 1997; McNevin and Barford 2000; Goldstein 1996, 1999; Singhal et al. 1996; Janni et al. 1998, 2001; Naylor et al. 1988; Kim et al. 2002; Lehtomaki et al. 1992; Luo, and Lindsey 2006; Luo and Oostrom 1997; Dawson 1993; Hodge et al. 1992; Kiared et al. 1996, 1997; Pond 1999; Govind and Bishop 1996; Kennes et al. 2007). With biological waste treatment methods, reactor engineering is often less complicated and therefore costs are less. In addition, usually no secondary wastes are produced. Biological methods are nonhazardous and benign for the environment. Possible drawbacks are restricted knowledge about the biodegradation processes, limited process control, and comparatively slow reaction kinetics (Kosteltz et al. 1996). For efficient removal of pollutants, target pollutants have to be sufficiently biodegradable and bioavailable. A major advantage in the case of odour treatment is that biocatalysts have high affinity for the substrates, which allows efficient treatment of low influent concentrations. Biocatalysts also operate at room temperature and they have innocuous final products example carbon dioxide and water. Biological treatment to control odours has gone through a major development step. In the early days, the design and operation of the most applied system, a biofilter, was done mainly by trial and error. But now much progress has been made in many areas such as microbiology, process modelling, reactor design and reactor operation (Kennes and Veiga 2001, Shareefdeen and Singh 2005). The conventional biofilter has changed from being a black box to better-defined biological systems with better control of the biological treatment process. Several reactor configurations have been developed for different applications. The biological methods for waste gas treatment are traditionally classified as biofilters, biotrickling filters and bioscrubbers. In the wastewater industry, biotrickling filter-type systems are currently the most common, but are often also referred to as bioscrubbers. But besides biotrickling filter and biofilters, injection of odourous air into the aeration tank of a wastewater treatment plant is used for treatment of low airflows with typically high-strength odours. Conventional biofilters use many types of organic material as support for the microorganisms and sometimes the biofilter media is mixed with granular, inorganic materials (e.g., lava rocks, clay balls or perlite) to stabilize the structure and to prevent preferential air flows and to increase the life of the media. Conventional biofilters using organic or partly organic media are used less often, as they face operating stability problems and important design limitations. The odourous air from wastewater treatment systems usually contain hydrogen sulphide, which is oxidised to sulphuric acid in a biofilter system. Table 1.5 shows a summary of important improvements that have been made on biological odour control technology in the recent past. All these improvements have resulted in systems which are now more reliable and simpler to operate and require less expense.

Table 1.5 Improvements of biological odour control system in the recent past. (Based on Kraakman n.d.)

Improvements	Advantages
The introduction of a mobile water phase	A better control of important microbial parameters
The use of synthetic support	A larger void volume
	More stable long term operation
	Lower pressure drop
Improved water and air distribution	A smaller reactor systems and lower outlet concentrations
Phase separation of gas transport and biological activity	A lower pressure drop and/or taller reactor systems
Multi-layer approach	A better use of biological capacities and lower outlet odour concentrations

References

AMEC, University of Calgary (2004) Assessment report on reduced sulphur compounds for developing ambient air quality objectives. http://environment.gov.ab.ca/info/library/6664.pdf. Accessed 11 March 2013

Anderson K (1970) Formation of organic sulphur compounds during Kraft pulping. II. Influence of some cooking variables on the formation of organic sulphur compounds during kraft pulping of Pine. Svensk Paperstid 73(1):1

Andersson B, Lovblod R, Grennbelt P (1973) Diffuse emissions of odourous sulfur compounds from kraft pulp mills, 1 VLB145, Swedish Water and Air Pollution Research Laboratory, Gotenborg

Anit S, Artuz R (2000) Biofiltration of air. Rensselaer Institute. www.rensselaer.edu/dept/chem-eng/Biotech-Environ/MIS/biofilt/biofiltration.htm. Accessed 15 Nov 2013

ATSDR (Agency for Toxic Substances and Drug Registry) (2011) Medical management guidelines for hydrogen sulfide. http://www.atsdr.cdc.gov/mmg/mmg.asp?id=385&tid=67. Accessed 25 March 2014

ATSDR (Agency for Toxic Substances and Drug Registry) (2006) Toxicological profile for hydrogen sulfide. http://www.atsdr.cdc.gov/toxprofiles/tp114.pdf. Accessed 20 March 2014

Bajpai P (2008) Chemical recovery in pulp and paper making. PIRA International, Leatherhead, 166 pp

BC Ministry of Environment (2009) Air quality objectives and standards. http://www.bcairquality.ca/reports/pdfs/aqotable.pdf. Accessed 25 Jan 2013

Biermann CJ (1996) Handbook of pulping and papermaking, 2nd edn. Academic, New York

Brauer H (1986) Biological purification of waste gases. Int Chem Eng 26(3):387–395

Burgess JE, Parsons SA, Stuetz RM (2001 Feb) Developments in odour control and waste gas treatment biotechnology: a review. Biotechnol Adv 19(1):35–63

Campagna D, Kathman SJ, Pierson R, Inserra SG, Phifer BL, Middleton DC (2004 Mar) Ambient hydrogen sulfide, total reduced sulfur, and hospital visits for respiratory diseases in northeast Nebraska, 1998-2000. J Expo Anal Environ Epidemiol 14(2):180–187

Canadian Centre for Occupational Health and Safety (2012) Hydrogen sulfide. http://www.ccohs.ca/products/databases/samples/cheminfo.html. Accessed 20 March 2013

Chan AA (2006) Attempted biofiltration of reduced sulphur compounds from a pulp and paper mill in northern Sweden. Environ Prog 25(2):152–160

Chung YC (2007) Evaluation of gas removal and bacterial community diversity in a biofilter developed to treat composting exhaust gases. J Hazard Mater 144:377–385

Cloirec PL, Humeau P, Ramirez-Lopez EM (2001) Biotreatments of odours and performances of a biofilter and a bioscrubber. Water Sci Tech 44(9):219–226

Dawson DS (1993) Biological treatment of gaseous emissions. Wat Env Res 65(4):368–371

Devinny JS, Deshusses MA, Webster TS (1999) Biofiltration for air pollution control. Lewis Publishers, Boca Raton

Deshusses MA (1997) Biological waste air treatment in biofilters. Curr Opin Biotechnol 8:335–339

Farmer RW (1994) Biofiltration: process variables and optimization studies. MS Thesis, University of Minnesota (December 1994)

Finn L, Spencer R (1997, Jan) Managing biofilters for consistent odor and VOC treatment. Biocycle 01:1997

Goldstein N (1996) Odor control experiences: lessons from the biofilter. Biocycle 37(4):70

Goldstein N (1999) Longer life for biofilters. Biocycle 40(7):62

Govind R, Bishop DF (1996) Overview of air biofiltration—basic technology, economics and integration with other control technologies for effective treatment of air toxics. Emerging solutions VOC air toxics control. Proc Spec Conf (Pittsburgh, PA): 324–350

Gullichsen J (2000) Fiber line operations. In: Gullichsen J, Fogelholm C-J (eds) Chemical pulping—papermaking science and technology, Book 6A. Fapet Oy, Helsinki, p A19

Haahtela T, Marttila O, Vilkka V, Jappinen P, Jaakkola JJ (1992 Apr) The South Karelia Air Pollution Study: acute health effects of malodorous sulfur air pollutants released by a pulp mill. Am J Public Health 82(4):603–605

Hessel PA, Herbert FA, Melenka LS, Yoshida K, Nakaza M (1997 May) Lung health in relation to hydrogen sulfide exposure in oil and gas workers in Alberta, Canada. Am J Ind Med 31(5):554–557

Hodge DS, Medina VF, Wang Y, Devinny JS (1992) Biofiltration: application for VOC emission control. In Wukasch RF (ed) Proceedings of the 47th Industrial Waste Conference. Purdue University, West Lafayette, pp 609–619

Hort C, Gracy S, Platel V, Moynault L (2009) Evaluation of sewage sludge and yard waste compost as a biofilter media for the removal of ammonia and volatile organic sulfur compounds (VOSCs). Chem Eng J 152:44–53

Inserra SG, Phifer BL, Anger WK, Lewin M, Hilsdon R, White MC (2004) Neurobehavioral evaluation for a community with chronic exposure to hydrogen sulfide gas. Environ Res 95(1):53–61

Janni KA, Nicolai RE, Jacobson LD, Schmidt DR (1998) Low cost biofilters for odor control in Minnesota. Final Report 14 Aug 1998. Biosystems and Agricultural Engineering Department, University of Minnesota, St. Paul

Janni KA, Maier WJ, Kuehn TH, Yang CH, Bridges BB, Vesley D (2001) Evaluation of biofiltration of air, an innovation air pollution control technology. ASHRAE Trans 107(1):198–214

Kiared K, Bibeau L, Brzenzinski R, Viel G, Heitz M (1996) Biological elimination of VOCs in biofilter. Environ Prog 15(3):148–152

Kiared K, Wu G, Beerli M, Rothenbuhler M, Heitz M (1997) Application of biofiltration to the control of VOC emissions. Environ Technol 18(1):55

Kennes C, Veiga MC (eds) (2001) Bioreactors for waste gas treatment. Kluwer Academic Publishers, Dordrecht (ISBN: 0-7923-7190-9)

Kennes C, Veiga, MC, Yaomin J (2007) Co-treatment of hydrogen sulfide and methanol in a single-stage biotrickling filter under acidic conditions. Chemosphere 68(6):1186–1193

Kim H, Kim YJ, Chung JS (2002) Long term operation of a biofilter for simultaneous removal of H2 S and NH3. J Air Waste Mgmt Assoc 52(12):1389–1398

Kosteltz AM, Finkelstein A, Sears G (1996) What are the 'real opportunities' in biological gas cleaning for North America? Proceedings of the 89th Annual Meeting & Exhibition of Air & Waste Management Association, A & WMA, Pittsburgh, PA; 96-RA87B.02

Kraakman NJR (n.d.) Biological Odour Control at Wastewater Treatment Facilities - the present and the future (let's forget the past). www.awa.asn.au/uploadedfiles/Biological_odour_Control_Krakman_paper.pdf

Legator MS, Singleton CR, Morris DL, Philips DL (2001 Mar–Apr) Health effects from chronic low-level exposure to hydrogen sulfide. Arch Environ Health 56(2):123–131

Lehtomaki J, Torronen M, Laukkarinen A (1992) A feasibility study of biological waste air purification in a cold climate. In: Dragt AJ, van Ham J (eds) Biotechniques for air pollution abatement and odour control policies. Elsevier Science, Amsterdam, pp 131–134

Luo J, Lindsey S (2006) The use of pine bark and natural zeolite as biofitlter media to remove animal rendering process odours. Bioresouce Tech 97(13):1461–1469

Luo J, Oostrom A (1997) Biofilters for controlling animal rendering odour: a pilot scale study. Pure Applied Chem 69(11):2403–2410

Marttila O, Jaakkola JJ, Partti-Pellinen K, Vilkka V, Haahtela T (1995 Nov) South Karelia Air Pollution Study: daily symptom intensity in relation to exposure levels of malodorous sulfur compounds from pulp mills. Environ Res 71(2):122–127

McNevin D, Barford J (2000) Biofiltration as an odour abatement strategy. Biochem Eng J 5:231–242

Nanda S, Sarangi PK, Abraham J (2012 Feb) Microbial biofiltration technology for odour abatement: an introductory review. J Soil Sci Environ Manag 3(2):28-35 (http://www.academic-journals.org/JSSEM. doi: 10.5897/JSSEM11.090. ISSN 2141–2391). Accessed 22 April 2013

Naylor LM, Kuter GA, Gormsen PJ (1988) Biofilters for odor control: the scientific basis. Compost Facts. Glastonbury, International Process System, Inc

Ontario Ministry of the Environment (2007) Ontario air standard for total reduced sulphur. http://www.ontla.on.ca/library/repository/mon/20000/277839.pdf. Accessed 17 Feb 2013

Ottengraf SPP (1987) Biological system for waste gas elimination. Trends Biotechnol 5:132–136

Ottengraf SPP (1986) Exhaust gas purification (chap 12). In: Schonborn W, Rehm H-J, Reed G (eds) Microbial Degradations. Biotechnology, vol 8. VCH, Weinheim, 425–452

Ottengraph SPP, Van Denoever AHC (1983) Kinetics of organic compound removal from waste gases with a biological filter. Biotechnol Bioeng 25:3089–3102

Ottengraf SPP, Meesters JPP, van den Oever AHC, Rozema HR (1986) Biological elimination of volatile xenobiotic compounds in biofilters. Bioprocess Eng 1:61–69

Partti-Pellinen K, Marttila O, Vilkka V, Jaakkola JJ, Jappinen P, Haahtela T (1996 Jul-Aug) The South Karelia Air Pollution Study: effects of low-level exposure to malodorous sulfur compounds on symptoms. Arch Environ Health 51(4):315–320

Pond RL (1999) Biofiltration to reduce VOC and HAP emissions in the board industry. Tappi J 82(8):137–140

Rappert S, Muller R (2005) Microbial degradation of selected odourous substances. Waste Manag 25:940–954.

Shults WT, Fountain EN, Lynch EC (1970) Irreversible coma and hemolytic anemia following inhalation. J Am Med Assoc 211:2153–2154

Shareefdeen Z, Singh A (2005) Biotechnology for odour and air pollution control. Springer, Heidelberg

Smet E, Lens P, Van Langenhove H (1998) Treatment of waste gases contaminated with odorous sulfur compounds. Critical Reviews in Environmental Science Technology 28(1):89–117

Smook GA (1992) Handbook for Pulp & Paper Technologists, 2nd ed. Angus Wilde Publications, Vancouver.

Soccol CR, Woiciechowski1 AL, Vandenberghel LPS, Soares M, Neto GN, Thomaz Soccol V (2003 July) Biofiltration: an emerging technology. Indian J Biotechnol 2(3):396–410

Singhal V, Singla R, Walia AS, Jain SC (1996) Biofiltration—an innovative air pollution control technology for H_2S emissions. Chem Eng World 31(9):117–124

Slaughter JC, Lumley T, Sheppard L, Koenig JQ, Shapiro GG (2003) Effects of ambient air pollution on symptom severity and medication use in children with asthma. Annal Allergy Asthma Immunol 91(4):346–353

Springer AM, Courtney FE (1993) Air pollution: a problem without boundaries. In: Springer AM (ed) Industrial environmental control pulp and paper industry, 2nd edn. Tappi Press, Atlanta, pp 525–533

Thorsvold BR (2011) Biological odor control systems: a review of current and emerging technologies and their applicability. nfo.ncsafewater.org/Shared Documents/Web Site Documents/Annual Conference/AC 2011 Papers/WW_T.pm_03.45_Thorsvold.pdf

US EPA (1979) Kraft Pulping-Control of TRS Emissions from Existing Mills, EPA-45012-78-003b

US EPA (2003) Toxicological review of hydrogen sulfide. http://www.epa.gov/iris/toxreviews/0061tr.pdf. Accessed 20 Dec 2013

Chapter 2
Emissions from Pulping

2.1 Kraft Pulping

Kraft pulping involves the cooking of wood chips at high temperature and pressure in white liquor, which is a solution of sodium sulphide and sodium hydroxide (Smook 1992; Biermann 1996; Adams et al. 1997). The white liquor dissolves the lignin that binds the cellulose fibres together. Two types of digester systems—batch and continuous—are used for kraft pulping. Most kraft pulping is done in batch digesters, although the more recent installations are of continuous digesters (Gullichsen 2000). In the batch digester, after completion of the cooking, the contents of the digester are transferred to an atmospheric tank usually referred to as a blow tank. The total contents of the blow tank are sent to pulp washers, where the spent cooking liquor is separated from the pulp. The pulp then proceeds through washing, which is done in various stages. The pulp is then bleached, pressed and dried into the finished product. The 'blow' of the digester does not apply to continuous digester systems. The balance of the kraft process is designed to recover the cooking chemicals and heat. Spent cooking liquor and the pulp wash water are combined to form a weak black liquor which is concentrated in a multiple-effect evaporator system to about 55 % solids. Then it is concentrated to 65 % solids in a direct-contact evaporator by bringing the liquor into contact with the flue gases from the recovery furnace, or in an indirect-contact concentrator. The strong black liquor is then fired in a recovery furnace (Bajpai 2008; Smook 1992). Combustion of the black liquor provides heat which is used for producing process steam and also for converting sodium sulphate to sodium sulphide. Inorganic chemicals present in the black liquor are collected as a molten smelt at the bottom of the furnace. The smelt is dissolved in water to produce green liquor, which is transferred to a causticizing tank. In the causticizing tank, calcium oxide is added to convert the solution back to white liquor for return to the digester system. From the causticizing tank, lime mud is precipitated, which is calcined in a lime kiln to regenerate calcium oxide (Adams et al. 1997). For process heating, for driving equipment, for providing electric power, etc., many mills need more steam than can be provided by the recovery furnace alone. So, conventional industrial boilers that burn natural gas, coal, oil, bark and wood are most commonly used.

P. Bajpai, *Biological Odour Treatment,* SpringerBriefs in Environmental Science,
DOI 10.1007/978-3-319-07539-6_2, © The Author(s) 2014

The problem of kraft mill odour originating from the sulphide in the white liquor in the initial pulping has long been an environmental and public relations issue for the pulp and paper industry (Smook 1992; Springer and Courtney 1993). The kraft mill odour is caused predominantly by malodourous reduced sulphur compounds, or total reduced sulphur (TRS), namely, methyl mercaptan, dimethyl sulphide and dimethyl disulphide and hydrogen sulphide. Methyl mercaptan, dimethyl sulphide and dimethyl disulphide are the main volatile organic sulphur compounds and are formed in the pulping process, while hydrogen sulphide is formed in the recovery furnaces (US EPA 1973, 1976, 1986, 1993a, b, 2001; Das and Jain 2004; Pinkerton 1993, 1998, 2000a, b; Bordado and Gomes 1997, 2003; Someshwar and Pinkerton 1992; Anderson 1970). Reduction of odourous gas emissions in kraft mills will significantly improve the environmental competitiveness of the pulp and paper industry, and will also improve public relations with their respective surrounding communities. When it is more economically feasible, odour reduction, instead of odour elimination, can improve significantly the air quality and the environment of a kraft mill, since it will reduce the radius of the area being impacted by the odour emission.

Although significant reduction of TRS emission has been achieved in the pulp and paper industry in the last decade with advanced odour abatement technologies, subjective odour nuisance at very low concentrations still causes odour problems in the communities surrounding kraft mills. TRS formation in kraft pulping was studied as early as the 1950s, 1960s, and 1970s (Kringstad et al. 1972; Tormund and Teder 1987; Zhao and Zhu 2004; Wag et al. 1995; Frederick et al. 1996; Tarpey 1995; Zhang et al. 1999). Much of this research effort has been devoted to quantification and kinetics (Wag et al. 1995; Frederick et al. 1996; Tarpey 1995) of the organic sulphur compound formation. The general formation mechanism of the TRS has been described by Frederick et al. (1996) and Jarvensivu et al. (1997). The formation of methyl mercaptan and dimethyl sulphide is through the reaction of mercaptide ion and the methoxyl groups present in the pulping liquor (Wag et al. 1995; Tarpey 1995). Dimethyl disulphide is not formed in the pulping process; rather it is formed through the oxidation of methyl mercaptan when black liquor is in contact with air (Wag et al. 1995; Frederick et al. 1996; Tarpey 1995). Hydrogen sulphide is not formed in the normal pulping pH conditions, but rather in the downstream processes where the pH of the streams are reduced below 10 through its dissociation from sodium sulphide (Zhao and Zhu 2004; Frederick et al. 1996). Other significant sources of hydrogen sulphide formation are lime mud reburning, black liquor pyrolysis, and molten smelt dissolution processes (Zhao and Zhu 2004; Wag et al. 1995; Frederick et al. 1996).

Typical characteristics of the gaseous emissions from kraft pulp mill are shown in Table 2.1. Overall, the three most important source of odour production are black liquor combustion, weak black liquor concentration and the digestion process. It can be seen that the source of the largest volume of potential emissions is the recovery furnace, followed closely by the digester blow gases and the washer hood vents. But, the most concentrated emissions come from the digester blow and relief gases. About 0.1–0.4 kg of TRS is emitted per ton of pulp at 5 ppm in the recovery

Table 2.1 Typical offgas characteristics of kraft pulp mill. (Based on data from Andersson et al. 1973; US EPA 1973)

Emission source	Offgas flow rate (m³/ton pulp)	Concentration (ppm by volume)			
		H_2S	CH_3SH	CH_3SCH_3	CH_3SSCH_3
Digester batch					
Blow gases	3–6000	0–1000	0–10,000	100–45,000	10–10,000
Relief gases	0.3–100	0–2000	10–5000	100–60,000	100–60,000
Digester, continuous	0.6–6	10–300	500–10,000	1500–7500	500–3000
Washer hood vent	1500–6000	0–5	0–5	0–15	0–3
Washer seal tank	300–1000	0–2	10–50	10–700	1–150
Evaporator hotwell	0.3–12	600–9000	300–3000	500–5000	500–6000
BLO tower exhaust	500–1500	0–10	0–25	10–500	2–95
Recovery furnace	6–000–12,000	(after direct-contact evaporator)			
		0–1500	0–200	0–100	2–95
Smelt dissolving tank	500–1000	0–75	0–2	0–4	0–3
Lime kiln exhaust	1000–1600	0–250	0–100	0–50	0–20
Lime slacker vent	12–30	0–20	0–1	0–1	0–1

Table 2.2 Odour threshold concentration of TRS pollutants. (Based on data from Springer and Courtney 1993)

Reduced sulphur compound	Odour threshold concentration (ppb)
Hydrogen sulphide (H_2S)	8–20
Methyl mercaptan (CH_3SH)	2.4
Dimethyl sulphide (CH_3SCH_3)	1.2
Dimethyl disulphide (CH_3SSCH_3)	15.5

boiler flue gases. The main difficulty with TRS emission is their nauseous odour, which are detected by the human nose at very low concentrations. The odour threshold (odour detectable by 50 % of the subjects) concentrations of the principal TRS compounds emitted by kraft mills, which are only few parts per billion by volume (Springer and Courtney 1993) are shown in Table 2.2. TRS is more of a nuisance than a serious health hazard at low concentrations. Thus, odour control is one of the main air pollution problems in a kraft mill.

Oxides of both sulphur and nitrogen are also emitted in varying quantities from a few points in the kraft system. The main source of sulphur dioxide emission is the recovery furnace due to the presence of sulphur in the spent liquor used as a fuel. Sulphur trioxide is sometimes emitted when fuel oil is used as an auxiliary fuel. The lime kiln and smelt dissolving tank also emit some sulphur dioxide. The emission of nitrogen oxides is more general because nitric oxide is formed whenever oxygen and nitrogen, which are both present in air, are exposed to high temperatures. A small part of the nitric oxide formed may further oxidize to nitrogen dioxide. These two compounds, nitric oxide and nitrogen dioxide, are termed the total oxide of nitrogen. Under normal operating conditions, the temperature in the recovery furnace is not high enough to form large quantities of oxides of nitrogen (NO_x). The main source of NO_x emissions is the lime kiln. SO_x and NO_x emission rates from various kraft mill sources are shown in Table 2.3. Due to the variations in operating conditions at different mills, there are large variations in the emission rates. Large

Table 2.3 Typical emissions of SO_x and NO_x from kraft pulp mill combustion sources. (Based on data from US EPA 1973; Someshwar 1989)

Emission source	Concentration (ppm by volume)			Emission rate (kg/ton[a])		
	SO_2	SO_3	NO_x (as NO_2)	SO_2	SO_3	NO_x (as NO_2)
Recovery furnace						
No auxiliary fuel	0–1200	0–100	10–70	0–40	0–4	0.7–5
Auxiliary fuel added	0–1500	0–150	50–400	0–50	0–6	1.2–10
Lime kiln exhaust	0–200	–	100–260	0–1.4	–	10–25
Smelt-dissolving tank	0–100	–	–	0–0.2	–	–
Power boiler	–	–	161–232	–	–	5–10[b]

[a] kg/t of air-dried pulp
[b] kg/t of oil

amount of NO_x are generated when the flame temperature is above 1300 °C and oxygen concentration is higher than 2 %. Modern recovery boilers should have SO_x emissions below 100 ppm when properly operated. Sulphur emissions from power boilers are controlled by using fuels of low sulphur content.

Another type of odourous emissions of non-sulphur compounds is generated by the hydrocarbons associated with the extractive components of wood, such as terpenes and fatty and resin acids, and also from materials used in processing and converting operations, like defoamers, pitch control agents, bleach plant chemicals, etc. Compared to TRS emissions, these hydrocarbon emissions are small, but they may be odourous, or act as liquid aerosol carriers contaminated with TRS, or undergo photochemical reactions.

2.2 Emissions from Neutral Sulphite Semi-Chemical (NSSC) Pulping

The semi-chemical pulping process uses a combination of chemical and mechanical energy to extract pulp fibres. Wood chips are first partially softened in a digester with chemicals, steam and heat. After the chips are softened, mechanical methods complete the pulping process. After digestion, the pulp is washed to remove cooking liquor chemicals and organic compounds dissolved from the wood chips. Then this virgin pulp is mixed with 20–35 % recovered fibre or repulped secondary fibre to enhance machinability. The chemical portion of the pulping process—cooking liquors, process equipment—and the pulp washing steps are very similar to the kraft and sulphite processes. Presently in the mills, the chemical portion of the semi-chemical pulping process uses either a non-sulphur or neutral sulphite semi-chemical (NSSC) process (Biermann 1996). The NSSC process uses a sodium-based sulphite cooking liquor and the non-sulphur process uses either sodium carbonate only or mixtures of sodium carbonate and sodium hydroxide for cooking the wood chips (EPA 2001a, b).

Generally, the emissions from NSSC are much lower in comparison to those from the kraft process. As no sodium sulphide is present in the pulping liquor, both methyl mercaptan and dimethyl sulphide are absent from the gaseous emissions, a very low amount of reduced sulphur is emitted (Dallons 1979). The sulphur emissions from the sodium carbonate (sulphur-free) process has been traced to sulphur in the fuel oil and process water streams used. The emissions of sulphur dioxide and NO_x are similar to those of a kraft mill.

2.3 Emissions from Sulphite Pulping

The cooking liquor in the sulphite pulping process is an acidic mixture of sulphurous acid and bisulphite ion (Smook 1992). In preparing sulphite cooking liquors, cooled sulphur dioxide gas is absorbed in water containing one of four chemical bases—magnesium, ammonia, sodium or calcium. This process uses the acid solution in the cooking liquor to degrade the lignin bonds between wood fibres. Sulphite pulps can be bleached more easily and have less colour than kraft pulps, but are not as strong as kraft pulps. The efficiency and effectiveness of the sulphite process is also dependent on the type of raw material and the absence of bark. For these reasons, the use of sulphite pulping is not very common and has reduced significantly in comparison to kraft pulping over time (EPA 2001a, b).

The sulphite process mainly operates with acidic sulphur dioxide solutions and as a result sulphur dioxide is the principal emission. Organic reduced sulphur (RS) compounds are not produced if proper conditions are maintained in the process. Because the odour threshold is about 1000 times higher for sulphur dioxide than for RS compounds, sulphite mills generally do not face the odour problem of a kraft mill. Volatile compounds such as methyl mercaptan and dimethyl sulphide are not produced in sulphite pulping. The method of attack on lignin by sulphite liquor is quite different than that by kraft liquor. The sulphite process involves sulphonation, acid hydrolysis and acid condensation reactions (Rydholm 1965).

Typical emissions in the sulphite process are sulphur dioxide with special oxides of nitrogen (problems arising in the ammonium-base process). Sulphur dioxide is also emitted during sulphite liquor preparation and recovery. Very little sulphur dioxide emission occurs with continuous digesters. However, batch digesters have the potential for releasing large quantities of sulphur dioxide, depending on how the digester is emptied. Digester and blow-pit emissions in the sulphite process vary depending on the type of system in operation. These areas have the potential for being a major source of sulphur dioxide emission. Pulp washers and multiple-effect evaporators also emit sulphur dioxide.

2.4 Mechanical Pulping

In mechanical pulping, pulp fibres are separated from the raw materials by physical energy such as grinding or shredding, although some mechanical processes use thermal and/or chemical energy to pretreat raw materials (Smook 1992;

Table 2.4 Volatile organic carbon emission from TMP mill before treatment. (Based on Nordic Council of Ministers 1993)

Process stage	
Sparkling washer	Total organic carbon: 6000 mg/m^3 (highest individual value: 9600 mg/m^3)
	Pinenes 1): 13,000 mg/m^3
Washing of woodchips	Total organic carbon: 300 mg/m^3
	Pinenes 1): 500 mg/m^3
Evacuation of air from other chests	Total organic carbon: 150 mg/m^3
	Pinenes 1): 50 mg/m^3

Biermann 1996). The main processes are stone groundwood pulping (SGW), pressure groundwood pulping (PGW), thermomechanical pulping (TMP) or chemithermomechanical pulping (CTMP). Emissions to the air are modest in mechanical pulping. Production emissions of purchased electricity can be high (Nordic Council of Ministers 1993). Atmospheric emissions from mechanical pulping are mainly linked to emissions of volatile organic compounds. Sources of volatile organic compounds emissions are evacuation of air from woodchips washing chests and other chests, and from sparkling washer where steam released in mechanical pulping processes contaminated with volatile wood components is condensed. The concentrations of volatile organic compounds depend on the quality and freshness of the raw material and the techniques applied. The emitted substances include acetic acids, formic acids, ethanol, pinenes and turpenes. Emissions of volatile organic compounds from a TMP mill before treatment is shown in Table 2.4

There are different alternatives for reducing volatile organic compounds emissions. Recovery of turpenes from those contaminated condensates that contain mainly turpenes or incineration of the exhaust gas in the on-site power plant or a separate furnace are available alternatives. In that case, about 1 kg volatile organic compounds/t of pulp is emitted from the process. Some volatile organic compounds may be released from wastewater treatment and unquantified emissions also occur from chip heaps.

In a CTMP mill, the atmospheric emissions originate mainly from chip impregnation and steam recovery (volatile organic compounds) and the bark boilers where wood residuals are burned (particulates, sulphur dioxide, NO$_x$) (European Commission 2001). As in other pulp and paper mills, mechanical pulping generates emissions to the air that are not process related but mainly related to energy generation by combustion of different types of fossil fuels or renewable wood residuals. The fossil fuels used are coal, bark, oil and natural gas. In a typical integrated paper mill that uses mechanical pulp high-pressure steam is generated in a power plant. The energy is partially transformed into electricity in a back pressure turbo generator and the rest is used in paper drying. The power plants burning solid fuels have electrostatic precipitators for the removal of particulates from the flue gases. The emission of sulphur dioxide is limited by using selected fuels. Depending on the local conditions there are paper mills using different amounts of energy from external supply.

References

Adams TN, Frederik WJ, Grace TM (1997) Kraft recovery boilers. TAPPI, Atlanta (USA 99)

Anderson K (1970) Formation of organic sulphur compounds during Kraft pulping. II. Influence of some cooking variables on the formation of organic sulphur compounds during kraft pulping of Pine. Svensk Paperstid 73(1):1

Andersson B, Lovblod R, Grennbelt P (1973) Diffuse emissions of odourous sulphur compounds from kraft pulp mills, 1 VLB145. Swedish Water and Air Pollution Research Laboratory, Gotenborg

Bajpai P (2008) Chemical recovery in pulp and paper making. PIRA International, U.K. 166 p

Biermann CJ (1996) Handbook of pulping and papermaking, 2nd edn. Academic, New York

Bordado JCM, Gomes JFP (1997) Pollutant atmospheric emissions from Portuguese Kraft pulp mills. Sci Total Environ 208(1–2):139–143

Bordado JCM, Gomes JFP (2003) Emission and odour control in kraft pulp mills. J Clean Prod 11:797–801

Dallons V (1979) Multimedia assessment of pollution potentials of non-sulphur chemical pulping technology. Environmental Protection Agency, Office of Research and Development, Industrial Environmental Research Laboratory, Cincinnati (EPA-600/2–79-026, January 1979)

Das TK, Jain AK (2004) Pollution prevention advances in pulp and paper processing. Environ Prog 20(2):87–92

EPA (2001a) Pulp and paper combustion sources National Emission Standards for Hazardous. U.S. Environmental Protection Agency, Office of Air and Radiation, Office of Air Quality Planning and Standards, Washington, DC

EPA (2001b) Pulping and bleaching system NESHAP for the pulp and paper Industry: a plain English description. U.S. Environmental Protection Agency. EPA-456/R-01–002. September 2001.http://www.epa.gov/ttn/atw/pulp/guidance.pdf. Accessed 12 Nov 2012

European Commission (2001) Integrated Pollution Prevention and Control (IPPC). Reference document on best available techniques in the pulp and paper industry. Institute for Prospective Technological Studies, Seville

Frederick WJ, Danko JP, Ayers RJ (1996) Control of TRS emissions from dissolving-tank vent stacks. TAPPI J 79(6):144

Gullichsen J (2000) Fibre line operations. In: Gullichsen J, Fogelholm C-J (eds) Chemical pulping—papermaking science and technology. Fapet Oy, Helsinki, p. A19 (Book 6 A)

Jarvensivu M, Lammi R, Kivivasara J (1997) Proceedings of the 1997 TAPPI Int. Environmental Conference, 645

Kringstad KP, McKean WT, Libert J, Kleppe PJ, Laishong C (1972) Odour reduction by in-digester oxidation of kraft black liquor with oxygen. TAPPI J 55(10):1528

Nordic Council of Ministers (1993) Study on Nordic pulp and paper industry and the environment. Nordic Council of Ministers, 1993

Pinkerton JE (1993) Emissions of SO_2 and NO_x from pulp and paper mills. J Air Waste Manag Assoc 43:1404–1407

Pinkerton J (1998) Trends in U.S. pulp and paper mill S and N emissions, 1980–1995. TAPPI J 181:114–122

Pinkerton J (2000a) Sulphur dioxide and nitrogen oxides emissions from pulp and paper mills in 2000. Special Report No. 02–06. National Council for Air and Stream Improvement, Inc., Research Triangle Park, NC, 2002

Pinkerton JE (2000b) Pulp and paper air pollution problems, Industrial environmental control. Pulp and paper industry. In: Springer AM (ed) 3rd edn. Atlanta, TAPPI, pp 501–535, 711 pp (Chap. 26)

Rydholm SA (1965) Pulping process. Wiley, New York, p 452

Smook GA (1992) Handbook for pulp & paper technologists, 2nd edn. Angus Wilde Publications, Vancouver

Someshwar AV (1989) Impact of burning oil as auxiliary fuel in kraft recovery furnaces upon SO_2 emissions. NCASI Technical Bulletin No. 578, December 1989

Someshwar AV, Pinkerton JE (1992) Wood processing industry. In: Buonicore AD, Davis WT (eds) Air pollution engineering manual. Van Nostrand Reinhold, New York

Springer AM, Courtney FE (1993) Air Pollution: a problem without boundaries. In: Springer AM (ed) Industrial environmental control pulp and paper industry, 2nd edn. TAPPI, Atlanta, pp 525–533

Tarpey T (1995) Odour reduction at [Daishowa- Marubeni International Ltd.'s] Peace River Pulp [Division Peace River Alberta]. International Environmental Conference Proceedings, 589

Tormund D, Teder A (1987) Elimination of malodourous organic sulphur compounds from the kraft pulping process with polythionate and sulfite. Nordic Pulp Paper Res J 2(3):97

US EPA (1973) Atmospheric emissions from the pulp and paper manufacturing industry. EPA-450/1-73-002. USEPA, Research Triangle Park

US EPA (1976) Environmental pollution control pulp and paper industry, Part 1, Air, U.S. EPA Technology Transfer Series, EPA-625/7-76-001, October 1976

US EPA (1986) Compilation of air pollutant emission factors, vol I. USEPA, Research Triangle Park

US EPA (1993a) Pulp, paper and paperboard industry background information for proposed air emission standards, manufacturing processes at kraft, sulfite, soda, and semi-chemical mills. EPA-453 R-93 – 050a. Office of Air Quality Planning and Standards, Research Triangle Park, NC27711

US EPA (1993b) Development document for proposed effluent limitations guidelines and standards for the pulp, paper and paperboard point source category. EPA-821-R-93-019. Office of Water, Mail Code 4303, Washington, DC

US EPA (2001) Air pollutants: a plain English description. EPA-456/R-01-003. September 2001. http://www.epa.gov/ttn/atw/pulp/Chaps.1-6pdf.zip. Accessed 18 Nov 2012

Wag KJ, Frederick WJ, Sricharoenchaikul V, Grace TM, Kymalainen MW (1995) Sulfate reduction and carbon removal during kraft char burning. International Chemical Recovery Conference: Preprints B, B35–50

Zhang Z, Luan G, Du P, Guo H (1999) More suitable process for dealing with the malodourous gases from sulfate cooking. Chin Pulp Paper Ind (4):16–18

Zhao H, Zhu J (2004) Operation experience of Kvaerner NCG System. Paper Paper Mak (1):21–23

Chapter 3
Biological Methods for the Elimination of Odourous Compounds

Several methods are available for the removal of odourous components from gaseous emissions (Bajpai et al. 1999). These include: gas phase methods, liquid phase methods, solid phase methods, combustion methods, and biological methods (Ottengraf 1986, 1987; Ottengraf et al. 1986). Physical–chemical waste gas cleaning techniques have proven their efficiency and reliability and will continue to occupy their niche, but several disadvantages remain. Among them are high investment and operation costs and the possible generation of secondary waste streams.

Biological methods have been attracting an increasing popularity because of the following reasons:

- Low cost
- Operational simplicity
- Intrinsically "clean technologies" as they reduce or eliminate the need for additional treatment of end products

Biological methods have a broad spectrum of applications. They are regarded as the most competitive systems for the deodorization of waste gases characterized by high flow rates and low concentrations of contaminants (Thorsvold 2011; Brauer 1986; Cloirec et al. 2001; Deshusses 1997; Devinny et al. 1999; Farmer 1994; Finn and Spencer 1997; McNevin and Barford 2000; Goldstein 1996, 1999; Singhal et al. 1996; Janni et al. 1998, 2001; Naylor et al. 1988; Kim et al. 2002; Lehtomaki et al. 1992; Luo and Lindsey 2006; Luo and Oostrom 1997; Dawson 1993; Hodge et al. 1992; Kiared et al. 1996, 1997; Pond 1999; Rappert and Muller 2005). Moreover, biological treatment is environmentally safe as it does not produce any harmful compounds. It is generally operated at natural conditions (normal atmospheric temperature and pressure) with no or gentle modifications. Nevertheless, biological filtration, also known as biofiltration, is regarded as the best available control technology for treating odourous gases and a "green technology" as it does not use any chemicals or produce any wastes that are potentially dangerous for the environment (Hort et al. 2009; Chung 2007). Biological odour treatment systems utilize biochemical processes to break down odourous compounds. These systems have been around for more than 100 years. In the past three decades, the trend toward biological treatment of odours has increased rapidly. These methods generally

P. Bajpai, *Biological Odour Treatment,* SpringerBriefs in Environmental Science, 17
DOI 10.1007/978-3-319-07539-6_3, © The Author(s) 2014

have the specific advantage that the pollutants are converted to harmless or much less harmful oxidation products, e.g. carbon dioxide, water, etc. These processes do not generally give rise to new environmental problems, or if they do these problems are minimal. An exhaust air problem should preferably not become a solid waste or waste water problem. Another advantage of biological treatment is the possibility of carrying out the process at normal temperature and pressure. Moreover, the process is reliable and relatively cheap, while the process equipment is simple and generally easy to operate. The elimination of volatile compounds present in waste gases by microbial activity is due to the fact that these compounds can serve as an energy source and/or a carbon source for microbial metabolism. Therefore, a broad range of compounds of organic as well as of inorganic origins can be removed by biological processes. As microorganisms need a relatively high water activity, these reactions generally take place in the aqueous phase, and as a result, the compounds to be degraded as well as the oxygen required for their oxidation first have to be transferred from the gas phase to the liquid phase. Hence, mass transfer processes play an important role in this technique. The microbial population can either be freely dispersed in the water phase or is immobilized on a packing or carrier material.

There are two main types of bacterial processes in a biological filter: autotrophic and heterotrophic. Autotrophic bacteria break down inorganic reduced sulphur compounds, mainly hydrogen sulphide, whereas heterotrophic organisms break down organic reduced sulphur compounds—mercaptans, dimethyl disulphide, etc. The autotrophic process converts hydrogen sulphide to sulphuric acid and is much more rapid in comparison to the heterotrophic breakdown of long-chain organic reduced sulphur compounds. Water added to the top of the unit is used to flush the sulphuric acid waste product away from the autotrophic bacteria, and this reduces the pH of the media bed beneath it. Because of this rapid conversion and the production of an acidic waste product, stratification of the media occurs with an autotrophic bacterial layer forming bottom-up on the air inlet side of a biofilter/biotower. The depth of this layer is dependent on the concentration of hydrogen sulphide in the airstream as a heterotrophic layer will not form until all the hydrogen sulphide has been removed. This is because while autotrophs flourish at low pH, heterotrophs cannot survive on surfaces with a pH less than 6. The ability of a biofilter/biotower to remove organic reduced sulphur compounds depends on the bed depth, empty bed residence time, and hydrogen sulphide concentration that must be removed first. One of the issues with biological systems is the difficulty in removing organic reduced sulphur compounds to sufficiently low levels. Even after removal of more than 99.9 % of the hydrogen sulphide (which can be accomplished easily biologically), the remaining air stream can still be quite odourous due to the remaining organic reduced sulphur compounds that remain at fairly low concentrations.

Biological waste gas purification technology currently includes bioreactors (Kennes and Veiga 2001; Kraakman 2004, 2005) known as:

- Biofilters
- Biotrickling filters
- Bioscrubbers (Table 3.1)

Table 3.1 Bioreactors for waste gas treatment. (Based on Ottengraf (1987))

Reactor type	Microorganisms	Aqueous phase
Biofilter	Fixed	Stationary
Biotrickling filter	Fixed	Flowing
Bioscrubber	Suspended	Flowing

The modes of operation for all these reactors are very similar. Air containing volatile compounds is passed through the bioreactor where the volatile compounds are transferred from the gas phase into the liquid phase. Microorganisms, such as bacteria or fungi, grow in this liquid phase and are involved in the removal of the compounds acquired from the air. The microorganisms performing the biodegradation normally grow as a mixture of different organisms. Such a mixture of different bacteria, fungi, and protozoa depends on a number of interactions and is often referred to as a microbial community. Microorganisms are generally organized in thin layers called biofilms. In most cases the pollutants in the air such as toluene, methane, dichloromethane, ethanol, carboxylic acids, esters, aldehydes, etc. (Tolvanen et al. 1998) act as a source of carbon and energy for growth and maintenance of the microorganisms. It must be noted that some waste gases, such as those produced during composting, are composed of many different chemicals like alcohols, carbonyl compounds, terpenes, esters, organosulphur compounds, ethers, ammonia, hydrogen sulphide, and many others (Tolvanen et al. 1998; Smet et al. 1999).The amazing aspect of the microbial community is that it generally develops to a composition so that all these different chemicals are removed and metabolized concurrently. Microorganisms also need essential nutrients and growth factors in order to function and produce new cells. The essential nutrients include nitrogen, phosphorus, sulphur, vitamins, and trace elements. Very often, these nutrients and growth factors are not present in the waste gas and have to be supplied externally. There are important differences between the three types of reactors mentioned above. They range from the way microorganisms are organized, i.e. immobilized or dispersed to the state of the aqueous phase in the reactor, i.e. mobile or stationary. The latter significantly impacts the mass transfer properties of the system.

Biological treatment of contaminated air is always not suitable (Soccol et al. 2003). First, because biotechnological techniques for air cleaning are only efficient and cost-effective in treating large volumetric air streams with low level of pollutants, up to 1–5 g m^3 (van Groenestijn and Hesselink 1993). The process becomes more costly when concentrations are higher than 50 g m^3, because of the requirements of moisture and temperature control (Gerrard et al. 2000). Second, the microorganisms present must efficiently carry out the conversion of the contaminants into harmless compounds. Third, the air contaminated with oil, grease, and dust accumulates and clogs the filter bed. Finally, the degradation of contaminants is dependent also on its water solubility and on its biodegradability. Bohn (1992) has reported that biofiltration may not be suitable for highly halogenated compounds, such as trichloroethylene, trichloroethane, and carbon tetrachloride because of low aerobic degradation, which means longer residence times and larger bed volumes.

Table 3.2 Biofiltration history. (Based on Anit and Artuz (2000))

1923	Biological methods proposed to treat odourous emissions
1955	Biological methods applied to treat odourous emissions at low concentrations in Germany
1960s	Biofiltration used for treatment of gaseous pollutants both in Germany and the USA
1970s	Biofiltration used with high success in Germany
1980s	Biofiltration used for the treatment of toxic emissions and VOCs from industry
1990s	More than 500 biofilters operating both in Germany and the Netherlands and biofiltration spreading widely in the USA

Furthermore, the size of a biofilter is inversely proportional to the degradation rate (Chitwood et al. 1999).

A description of each of the three types of bioreactors for biological waste gas purification currently in use is given:

3.1 Biofilters

Biofilters started to gain popularity in the late 1980s at waste water treatment plants in the USA due to the low life-cycle costs and the better understanding of how to design and operate them successfully (Cáceres 2010; McNevin and Barford 2000; Burgess et al. 2001; Wani et al. 1997; Govind and Bishop 1996; Swanson and Loehr 1997). Biofiltration uses microorganisms that can oxidize many compounds, and thus has potential for being used for the abatement of odours, volatile organic compounds (VOCs), and air toxins (Kennes and Veiga 2001; Kennes et al. 2007; Ottengraf 1987). The concept of biofiltration is actually not new, it is an adaptation of the process by which the atmosphere is cleaned naturally (Bohn 1992). Biofiltration is the oldest biotechnological method for the removal of undesired off-gas components from air (Table 3.2). Since the 1920s, biofilters have been applied to remove odourous compounds from waste water treatment plants or intensive animal farming (van Groenestijn and Hesselink 1993). Earlier, they were made by digging trenches, laying an air distribution system, and refilling the trenches with permeable soil, wood chips, or compost. From the late 1970s, most of the development work on biological off-gas treatment has been carried out in Europe, especially in Germany and the Netherlands, in response to increasingly national regulatory requirements. Only up to the 1980s, intensive progress had started in Western Europe and the USA (Ottengraf et al. 1986), and since then, research on biofiltration have been focused also on the degradation of toxic volatile chemicals and on industrial applications using different supports, types of filters, and microorganisms. Table 3.3 shows industries using biofiltration in Europe. All of these sources typically release large volumes of off-gases that contain only low concentrations of the target organic compounds. Table 3.4 presents an abbreviated list of chemicals that can be treated by biofiltration

Table 3.3 Industries using biofiltration in Europe. (Based on Leson and Winer (1991))

Chemical operations	Coffee roasting
Composting facilities	Chemical storage
Coca roasting	Landfill gas extraction
Film coating	Fish frying
Slaughter houses	Investment foundries
Flavours and fragrances	Tobacco processing
Print shops	Pet food manufacturing
Waste oil recycling	Industrial and municipal waste water treatment plants

(Barnes et al. 1995; Barshter et al. 1993; Ergas et al. 1995; Hodge et al. 1991a, b; Morgenroth et al. 1996; Mueller 1998; Ottengraph and Van Denoever 1983).

The first biofilters were built in the USA in the 1960s (Gerrard et al. 2000). During the late 1980s to the late 1990s, approximately 30 large full-scale systems having about more than 100 m³ of filter material have been built for the control of VOCs, hazardous air pollutants, and odour (van Lith et al. 1997). Biofiltration has had more industrial success in Europe and Japan. Documented in 1953, the conventional type biofilters were used for treating odourous sewer gases at Long Beach, California, USA (Nanda et al. 2012). Currently with its applications globally implemented, biofiltration is an entrenched air pollution control technology in several European countries (Singhal et al. 1996). Nearly, 40 % of animal rendering plants in New Zealand employ biofilters (Luo and van Oostrom 1997). In the USA and Canada, biofilters have already found mass recognition in many pig production houses (Moreno et al. 2010; Deutsch 2006; Miller et al. 2004), aquaculture ponds (Rogers and Klemetson 1985), and other similar facilities. In Asian countries like India and China, extensive research on biofilters is leading to their large-scale implementation (Saravanan et al. 2010; Chung et al. 2001; Chung 2007; Arulneyam and Swaminathan 2005, 2003).

Biofiltration has been also applied to other easily biodegraded volatile compounds and more complex mixtures. Since the 1980s, significant research efforts have been expended in an attempt to extend the application to more recalcitrant compounds, such as chlorinated and sulphurous species, and to mixtures of compounds. Efforts have been also made to improve the packing material in terms of its nutrient composition, pore structure, and mechanical integrity These properties are very important in extending the life of the system. There is a huge variation in packing material used industrially. Biofilters also vary greatly in size. Fouhy (1992) reports that the treated quantity of gas varies from 300m³ h⁻¹ at a landfill site, to over 200,000 m³ h⁻¹ at an animal rendering facility. Increasing regulatory strictness with respect to air emissions, such as the Clean Air Act Amendments (1990) in the USA, the Air Quality Framework Directive, its daughter, and relevant national legislation in Europe, and the Canadian Environmental Protection Act (1999), is driving further research into all air pollution control technologies. Biofiltration technology is a promising method of removal of odour, VOCs, and air toxin from waste gas streams because of: low capital and operating costs, low energy requirements, and an absence of residual products requiring further treatment or disposal (Wani et al. 1997).

Table 3.4 Chemicals treatable by biofiltration. (Based on Barnes et al. 1995; Barshter et al. 1993; Ergas et al. 1995; Hodge et al. 1991; Morgenroth et al. 1996; Mueller 1996; Ottengraph and Van Denoever 1983)

Acetate
Acetone
Ammonia
Benzene
Butanol
Butylaldehyde
Butyl acetate
Carbon monoxide
Mono-, di-, tri-chloromethane
Diethylamine
Dimethyl disulphide
Dimethyl sulphide
Ethanol
Ethylbenzene
2-Ethylhexanol
Hexane
Hydrogen sulphide
Indole
Isopropyl alcohol
Methane
Methanol
Methyl ethyl ketone
Dimethyl sulfide
Ethanol
Ethylbenzene
2-Ethylhexanol
Hexane
Hydrogen sulphide
Indole
Isopropyl alcohol
Methane
Methanol
Methyl ethyl ketone

The suitability and cost-effectiveness of biofiltration for treating VOCs has led to increased acceptance and use by the industry. Biofilters have been evaluated for treatment of many compounds, using a variety of packing media and bed configurations (van Groenestijn and Hesselink 1993). Researchers have developed numerical models of the process and begun to consider microbial dynamics and characterization (Deshusses 1997). Some researchers have attempted modelling long-term performance (Song and Kinney 2002). Various operational strategies, such as nutrient supplementation, the use of thermophilic bacteria, and cometabolism have been also researched.

Biofiltration is similar to the biological treatment of waste water or in-situ bioremediation of contaminated soils and hazardous sludge (Rozich 1995). The acceptance of biofiltration has followed from biotechnological advances that provide an increasingly thorough knowledge of the system and how the process can be optimized not only to achieve high removal efficiencies with low energy consumption but importantly also to achieve these elimination efficiencies over long periods of time with minimal operator intervention and/or need for maintenance (Marsh 1994). VOC emissions have become an essential issue for industrial operators as a result of the implementation of the US 1990 Clean Air Act Amendments and similar regulations in Europe, and thus is a major driving force for the exploration of cost-effective control options. Biofiltration is a promising technology for processes that emit large off-gas volumes with relatively low concentrations of contaminants. With respect to the purification of polluted air, biofiltration is a commonly applied technique to odour abatement, where it is an established control method. It has also demonstrated some success in controlling VOCs.

Biofilters do exceptionally well in the removal of odouriferous compounds and in the removal of VOCs (Ottengraf 1986; Hirai et al. 1990; Deshusses and Hammer 1993; Leson and Winer 1991), mainly solvents, from air. The pollutants are completely biodegraded without the formation of aqueous effluents under optimum conditions. As gases pass through a biofilter, odourous compounds are removed by sorption (absorption/adsorption) and bio-oxidation (Williams and Miller 1992). The odourous gases either adsorb onto the surface of the biofilter medium and/or are absorbed into the moisture film on the biofilter particles. The sorbed compounds are then oxidized/degraded by the microorganisms. Final products from the complete bio-oxidation of the air contaminants are carbon dioxide, water, mineral salts, and microbial biomass. The removal of gaseous pollutants in a biofilter is the result of a complex combination of different physicochemical and biological phenomena.

Essentially, a biofilter is a layer of biologically active media usually of natural origin. The filter particles are typically soil, compost, peat, wood chips, tree bark, and heather (Sercu et al. 2006). Granular activated carbon and plastic material are also used. One kind or several combinations of particles have been used. The media should provide a large surface area, nutrients, and moisture for the microbial activity and adsorption/absorption of the odourous molecules (Sercu et al. 2006). The microflora for the degradation of odours is a part of the package. There is no continuous water phase. For obtaining better results, the addition of nutrients containing nitrogen and phosphorus should be considered. However, this will add some cost to the process. The presence of bulking inerts usually requires the addition of nutrients, particularly with high load regimes (Devinny et al. 1999). Sufficient porosity of around 0.50 is necessary for low pressure drop (power requirements). To build a conventional open-bed filter, in the early ages of the technique, a hole was excavated in the ground around 1.0 m deep and filled up with a bed of the selected media. Nowadays, synthetic material or concrete is used. Perforated piping or other systems are used for gas distribution under the bed. The waste air flow, combined with the void fraction, causes the residence time to be normally between 15 and 60 s, the time it takes for the odours to be absorbed and metabolized through the filter. Devinny et al. (1999) have reported surface loading rates of about 1.2 m^3 m^{-2} min^{-1}. Impermeability is required to avoid liquid leaching. For optimal long-term operation of biofilters, next to controlling the biofilter moisture content, precautions should be taken to avoid acidification if sulphur- or nitrogen-containing

Fig. 3.1 Conventional open
type biofilter. (Reproduced
with permission from Sercu
et al. (2006))

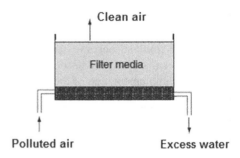

compounds are present. This can be achieved by buffering, e.g. by addition of calcium carbonate (Rafson 1998) or by regular replacement of the filter material every 1–5 years, depending on the loading rate. The latter treatment is also required to remove the end products or other accumulated intermediates, to avoid high pressure drops, and to prevent nutrient limitation if nutrients are not provided during operation of biofilter. The removal efficiencies range from 60 to 100 %, depending on the media and the pollutants contaminating the air. Initial performance is very good but as time goes on problems may arise which lead to severe loss of efficiency. Channelling and clogging are observed. Modular closed systems are commercially available (Sercu et al. 2006). These systems reduce the surface for installation because they are stacks of trays that can be set in series or parallel arrangements, or combinations of both. The usual time of operation, with good removal efficiency for conventional systems is extended due to selected media, uniform distribution, and the addition of controls for temperature, pH, moisture, and airstream relative humidity (Devinny et al. 1999; Rozich 1995; Marsh 1994; Bohn and Bohn 1988; Hodge et al. 1991; Ottengraf 1986, 1987; Bohn 1992, 1993).

Biofilters are generally constructed in a vessel packed with loose beds of solid material, soil, or compressed cakes with microorganisms attached to their surface. Waste gases are passed through these units via induced or forced draught. Biofilters are capable of handling rapid air flow rates and volatile organic carbon (VOCs) concentrations in excess of 1000 ppm. These units are gaining importance in bioremediation also and are timely in that they are a cost-effective means to deal with the more stringent regulations on VOC emission levels. There are mainly two types of biofilters:

Soil filter (open type biofilter)

The first and simplest is the soil filter. Contaminated air from a small waste stream or other treatment process is passed through a soil–compost type design, a so-called open system (Fig. 3.1). Sometimes, nutrients are preblended into the compost pile to provide conditions for growth of microorganisms and biodegradation of the waste by indigenous microorganisms. As they are usually installed in the open air and partly underground, these systems are exposed to many weather conditions such as rain, frost, temperature fluctuations, etc. These filters are normally overdesigned and require a very large area (Ottengraf 1986).

Closed type biofilter

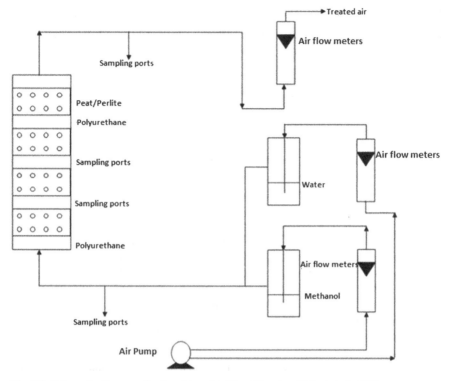

Fig. 3.2 Schematic diagram of a closed type biofilter. (Based on Bajpai et al. 1999)

Several closed type systems have been developed which have (one or more) treatment beds or discs of different packing materials or media, nutrients, microbial cultures, and/or compost in its reactor cell (Bajpai et al. 1999; Nanda et al. 2012; Shareefdeen et al. 1993). In the treatment bed, the waste air stream and the filter are humidified as the waste is passed through one, two, or more beds. In this approach, a series of humidified discs or beds are placed inside a reactor shell (Fig. 3.2). These layered discs contain packing material/media, nutrients, microbial cultures, and/or compost material. The waste air stream organics undergo biodegradation as they pass through the system. Any collected water condensate from the process is returned to the humidification system for reuse. Biofilters have reportedly been built to handle up to 3000 m^3 min^{-1} of air flow using filters up to 6500 m in wetted area (Anon 1991). The filters can be customized with specific carriers, nutrients blends, or microbial cultures. Some biofilters can endure up to 5 years before replacement is necessary (Holusha 1991). Spent filters can be utilized as fertilizer since they present no hazard. Multistage biofilters are one of the type that filters waste gas containing different components and require different conditions for microbial treatment. In recent years, there has been a lot of advancement in biofiltration technology that makes it easier to model a biological filtration process and ultimately design the suitable biofilter. Selection of the microbial culture for biofiltration is usually done

Table 3.5 Bacterial conversions in biofiltration. (Based on McNevin and Barford (2000))

Bioconversion	Nature of bacteria	Condition
Organic carbon oxidation	Chemoheterotrophic bacteria	Aerobic
Nitrification	Nitrifying bacteria	Aerobic
Sulphide oxidation	Sulphur-oxidizing bacteria	Aerobic
Denitrification	Denitrifying bacteria	Anaerobic

according to the composition of the waste air and the ability of the microorganism to degrade the pollutant present in it. Sometimes, a single microorganism is enough to degrade the pollutant and sometimes a consortium of microorganisms is used for catabolism. Some biological conversions occurring during biofiltration of odourous compounds are presented in Table 3.5.

There are a number of common problems encountered in conventional biofilter operation (Norman 2002). Maintaining proper moisture and nutrient content in the packing material is difficult. This can lead to system failure. Start up is also difficult, where a slow start-up period equates to an excessive period of contaminant breakthrough. One of the most common problems faced in full-scale implementation of biofilters is clogging. Clogging can cause channelling within the packing material, limiting the amount of contaminated air being treated (Devinny et al. 1999). Pressure drops, increasing wear, and energy demand on the system are associated with clogging. Clogging occurs when excess biomass collects in the void space of the packing material. Clogging is generally found at the biofilter's inlet due to biomass concentrations being highest in the area of greatest contaminant loading (Ergas et al. 1994). This interferes with the passage of the waste gas stream through the biofilter. Conventional biofilters are continuous flow processes designed and operated to receive a relatively constant stream of contaminated air (Irvine and Moe 2001). Such systems, usually designed for minimal operator control (often times only allowing adjustment of the system's moisture content), provide little opportunity for implementation of the engineering decisions which could increase the performance of biofilter during relatively steady-state conditions or transient periods of high contaminant loading. These transient conditions reflect the uncontrolled, unsteady-state conditions commonly encountered in most industrial processes.

Over the past few years, extensive research has been conducted on the microorganisms used in biofiltration. Diverse microbial communities such as bacteria, actinomycetes, and fungi are involved in biofiltration as they are indigenous to the biomedia such as soil and compost. Much of the research has been focused on bacteria but fungi have also been used in biofiltration (Spigno et al. 2003; Garcia-Pena et al. 2001; Cox et al. 1997). Compost has been found to house bacteria belonging to the groups Actinobacteria, Bacteroidetes, Firmicutes, and Proteobacteria (Chung 2007). Table 3.6 presents a few essential microorganisms and their consortium which are used in biofiltration to remove the odourous pollutants from waste gases. A good biofiltration always depends on its heterotrophic microbial population that use organic compounds as energy and carbon sources (Nanda et al. 2012). A major benefit of the biofiltration system is that the viability of the microorganisms is maintained for a longer period although the system is not in function for a longer period.

Table 3.6 Microorganisms used in biofiltration of waste gases. (Based on data from Leson and Winer 1991; van Lith et al. 1997; Bohn 1975; Pomeroy 1982, Hirai et al. 1990; Lee and Shoda 1989, Furusawa et al. 1984, Ottengraph et al. 1983; Sivela and Sundman 1975; Kanagawa and Kelly 1986; Kanagawa and Mikami 1989; Smith and Kelly 1988a, b; DeBont et al. 1981; Suylen et al. 1986; Suylen et al. 1987, Kirchner 1987; Shareefdeen et al. 1993)

Pollutants	Odour	Microorganisms
Hydrogen sulphide	Rotten eggs	*Bacillus cereus* var. mycoides, *Streptomyces* spp., *Hyphomicrobium* spp., *Thiobacillus* spp., *Thiobacillus thioparus* TK-m, *Xanthomonas* spp., *Methylophaga sulfidovorans*, *Pseudomonas putida*
Dimethyl sulphide	Decayed cabbage	*Pseudonocardia asaccharolytica* DSM 44247, *Hyphomicrobium* spp., *Thiobacillus* spp., *T. thioparus* TK-m, *Thiocapsa roseopersicina*, *P. putida* DS1
Dimethyl disulphide	Decayed cabbage	*P. asaccharolytica* DSM 44247, *Hyphomicrobium* spp., *Thiobacillus* spp., *T. thioparus* TK-m
Dimethyl trisulphide	Decayed cabbage	*P. asaccharolytica* DSM 44247
Carbon disulphide	Decayed pumpkin	*Paracoccus enitrificans*, *Thiobacillus* sp.
Methanethiol	Decayed cabbage	*Hyphomicrobium* spp., *Thiobacillus* spp., *T. thioparus* TK-m, *Arthrobacter* sp., *Bacillus* sp.
Dimethylamine	Putrid, fishy	*Hyphomicrobium* sp., *Methylobacterium* sp., *Psuedomonas aminovorans*, *Mycobacterium sp.*, *P. denitrificans*, *Methylophilus methylosporus*, *Micrococus* sp., *Pseudomonas* sp., *Paracoccus* sp. T231
Trimethylamine	Ammonical, fishy	*Aminobacter aminovorans*, *Paracoccus* sp. T231, *Paracoccus aminovorans*, *P. aminovorans*, *Hyphomicrobium* spp., *Micrococus* sp.
Diethylamine	Ammonical, fishy	*Pseudomonas citronellolis* RA1, *M. diernhoferi* RA2, *Hyphomicrobium* sp., *Pseudomonas* sp., *Candida utilis*, *Hansenula polymorpha*
Triethylamine	Ammonical, fishy	*P. citronellolis* RA1, *M. diernhoferi* RA2
Various volatile organic Carbons	Malodourous	*Actinomyces globisporus*, *Penicillum* sp., *Cephalosporium* sp., *Mucor* sp., *Micromonospora albus*, *Micrococcus albus*, *Ovularia* sp.

This is because of the use of natural materials as the filter bed (Ottengraph and Van Denoever 1983). Although limited information is available on the microbial communities involved in biofiltration of odourous compounds, new technologies such as denaturing gradient gel electrophoresis, temperature gradient gel electrophoresis, and single-strand conformation polymorphism have allowed better understanding of the microbial population dynamics in the natural and artificial systems (Xie et al. 2009; Chung 2007). Traditional microbiological methods have revealed the presence of mixed populations of bacteria, yeast, fungi, and higher organisms in the biofilters. Bacterial species of *Thiobacillus* and *Hyphomicrobium* degrade several sulphur compounds like hydrogen sulphide, methyl sulphide, dimethyl sulphide, dimethyl disulfide, dimethyl sulfoxide, methanethiol, etc. (Sivela and Sundman 1975; Kanagawa and Kelly 1986; Kanagawa and Mikami 1989; Smith and Kelly

1988a, b; DeBont et al. 1981; Suylen et al. 1986, 1987). For methanol bio-oxidation, *Pseudomonas fluorescens* and a bacterial consortium consisting of *Methylomonas, Aeromonas, Achromobacter, Flavobacterium, Alcaligenes,* and *Pseudomonas* have been used (Kirchner 1987; Shareefdeen et al. 1993).

The functioning of the biofilter is dependent on several factors; a few of them are discussed here. The nature of the filter medium is of great importance. The filter medium affects the growth of microorganisms and adsorption of pollutants as these pollutants have to be adsorbed on the filter medium to be available for biological transformations (Xie et al. 2009). For successful operation and avoid filter clogging, dusts, aerosols, grease, resins, accumulated products such as sulphate, etc. should be regularly removed by separators (Nanda et al. 2012). The waste gases occasionally contain some constituents that make microorganisms vulnerable to them but this can be avoided by installation of particulate filter before subjecting the gas to biofilter (Leson and Winer 1991). The concentration of pollutants and the loading rates also affect the performance of the biofilter. In order to increase oxygen concentration required by microorganisms packed in the carrier material, the air stream having high pollutant concentration should be diluted with fresh air (Yang and Allen 1994). Proper moisture level, usually 40–60 % and temperature 10–15 °C higher than ambient should be maintained within the biofilter (Nanda et al. 2012). The efficiency increases and vice versa with an increase in temperature. The optimum temperature range for removal of hydrogen sulphide is 35–50 °C (Nanda et al. 2012). Very often, there is a slow increase in temperature due to microbial respiration and exothermic reactions in the filter. Maintaining an optimum pH within the system exclusively depends on the microorganism used. Yang and Allen (1994) reported higher pH to increase the removal of hydrogen sulphide. However, sulphur-oxidizing bacteria have a wide pH range of 1–8. The addition of limestone can optimize the required pH range.

The selection of the biofilter media is very important for the performance of the process since all of the filter media allow polluted air to interact closely with degradative microorganisms, oxygen, and water (Bohn 1992; Schroeder 2002). The material used as filter medium must have some features that will be important for the performance of the biofilter. Physical media constitution is required to provide fine porous, huge surface area, and a uniform pore size distribution. Pore uniformity strongly defines the flow, and hence the effectiveness of the biofilter. The degree of porosity of the bedding material is essential to increase the adsorption of microorganisms on it. Fine or narrow pore size uniformly distributed all over the medium increases the uniformity of air and water flow through the bed (Bohn 1992). Inorganic bed material, containing a range of metal oxides, glass, or ceramic beads have good flow properties, because of its uniform shape. In case of inorganic material, the degree of porosity can be controlled in certain limits (Cohen 2001). Polyvinyl chloride is often used as a packing material, but because of its smooth surface, it took much more time to reach a maximum loading rate than when porous red clay and grey potter's clay is used (Van den Berg and Kennedy 1981). Besides, as the active biofilm will adhere onto the biofilter medium, the amount of microorganisms presented will be dependent on the available surface, increasing the efficiency of the biofilter. Generally, the packing material is a mixture of natural fibrous material having a large specific area and a coarse fraction. According to Van Groenestijn and Hesselink (1993) and Anit and Artuz (2000), an ideal biofilter medium should have large surface area for adsorption of contaminants as well as to support the growth of microorganisms. The material selected for biofilter medium should be physically stable and very well structured in order to ensure that the medium does not compact, get smaller, or accumulate during the time of operation. This will develop preferential flow in the bed and decrease the available medium area or cause filter plugging. Biofilter medium material should have physical properties like physical stability and ease of handling (Bohn 1996; Anit and Artuz 2000). If the medium is inert, it should have a large population of microorganisms on it, which will oxidize all organic compounds. Synthetic or inert media should be inoculated with soil, compost, or sewage sludge. These materials have a large and complex population of microorganisms available to develop the proper microbial culture for the process (Schroeder 2002; Bohn 1996). Pure culture can also be tested as inoculum. Other points to be assessed are operation and maintenance costs. Lifetime of the medium is also very important. The replacement of dirty, disruptive bed material is a costly affair. The old medium should be dumped, which is sometimes very expensive (Bohn 1996). An appropriate biofilter material medium must have the ability to retain moisture to sustain the biofilm layer. It should also have the capacity to retain nutrients and supply them to the active biofilm formed by microorganisms when needed (Anit and Artuz 2000). According to Schroeder (2002), the decomposed peat and overtime compost should be replaced every 3–5 years. During the operation period, maintenance is required to keep the biofilter effective. Plastic and ceramic media must be cleaned every few months to break up surfaces of biofilm, in order to avoid plugging. It is important to select materials that are able to keep hydrophilicity to avoid dryness because degradative microorganisms require moist conditions to thrive (Bohn 1996). Finally, the medium cost must be analysed to make the installation feasible. Low price and accessible material must be found, because the cost of the material and the transportation fees may make the treatment unviable. Many materials are available to be used as media in biofilters (Table 3.7; Carlson and Leiser 1966; van Lith and Leson 1997; Bohn and Bohn 1988; Bohn 1975; Pomeroy 1982; Lee and Shoda 1989; Furusawa 1984; Ottengraph and Van Denoever 1983; Sivela and Sundman 1975; Van Langenhove et al. 1986; Luo and van Oostrom 1997; Campbell and Connor 1997; Hodge and Devinny 1994; Eisenring 1997; Govind and Bishop 1996; Ottengraf 1986; Hirai et al. 1990; Leson and Winer 1991; Shareefdeen et al. 1993; Morton and Caballero 1997). Typical biofilter medium material includes compost-based material, earth, plastic, or wood-products based material. Table 3.8 shows the advantages and disadvantages of some Biofilter media. Biofilter medium provides a large surface area for the adsorption and absorption of contaminants and also serves to provide nutrients for the microbial population. For some types of media, lack of proper nutrients will require addition of nutrients, such as nitrogen or phosphorus compounds (Anit and Artuz 2000). Plastic packing, such as plastic rings of various sizes and porous diatomaceous earth pellets are tested in laboratory studies (Sorial et al. 1997; Wright et al. 2000). Mixtures of media types are sometimes used to provide operational advantage. Using a soil, peat, or compost bed, the medium can provide some or all essential nutrients required for the microbial growth. Depending

Table 3.7 Packing materials used for biofilter. (Based on Carlson and Leiser (1966); van Lith and Leson 1997; Bohn and Bohn 1988; Bohn 1975; Pomeroy 1982; Lee and Shoda 1989; Furusawa 1984; Ottengraph and Van Denoever AHC 1983; Sivela and Sundman 1975; Van Langenhove et al. 1986; Luo and van Oostrom 1997; Campbell and Connor 1997; Hodge and Devinny 1994; Eisenring 1997; Govind and Bishop 1996; Ottengraf 1986; Hirai et al. 1990; Leson and Winer 1991; Shareefdeen et al. 1993)

Activated carbon
Activated carbon fabric
Bark
Bioton
Compost
Granulated activated carbon + sintered diatomaceous earth
Peat
Peat + perlite
Polyurethane foam
Sintered diatomaceous earth
Soil
Structured ceramic media
Textile

on the requirements other agents can be added (Adler 2001). Normally the biofilter bed media is composed of a mixture of activated carbon, alumina, silica, and lime, alternatively, soil, peat, or more refined material, such as inert material, cellulous material, or mineral material, combined with a microbial population that enzymatically catalyses the oxidation of the absorbed, adsorbed, or dissolved gases. The most common material used as bed media is a mixture of compost and a inert charge to give support to the bed such as wood chips or bark, silica, perlite, or synthetic media, basically, polystyrene beads (Bohn 1992). Organic bedding material has a higher adsorptivity when compared to inorganic material. For example, microbial adsorption is 248 mg g^{-1} for wood chips and 2 mg g^{-1} for inorganic silica. This is attributed to the larger variety of reactive groups carboxyl, amino hydroxyl, etc., and the presence of certain quantity of nutrients presented on organic material that help the attachment of the microorganisms (Cohen 2001). Nutrients or a buffer for the gas treatment process can be added to the packing or can be extracted from natural packing such as compost, soil, or peat.

Several different aspects—isolation and characterization; the use of pure cultures of bacteria and fungi; mixed microbial populations; effect of enrichment culture including application of special strains, types of microorganisms and their metabolic activities; effects of external conditions on microbial activity and release of microorganisms from biofilters—have been studied regarding the microbiological potential of biofilters (Cho et al. 1991; Cox and Deshusses 1999; Cox et al. 1997; Shareefdeen et al. 1993; Mallakin and Ward 1996; Andreoni et al. 1997; Lipski et al. 1992; Reichert et al. 1997; Diks and Ottengraf 1994; Krishna et al. 2000; Woertz and Kinney 2000; Zilli et al. 1996; Weigner et al. 2001; Ottengraf and Konongs 1991; Becker and Rabe 1997; Bendinger et al. 1992).

The presence of microorganisms in the biofilter media has raised concern over their potential release into the treated off-gas and the consequential exposure to

Table 3.8 Advantages and disadvantages of various biofilter media

Soil

Advantages

Well-established technology

Suitable for low contaminant concentrations or odour control

Low-cost media

Disadvantages

Prone to channelling and maldistribution

Humidification needed

Limited ability to neutralize acidic degradation products

Low adsorption capacity

Eventual media replacement required

Low biodegradation capacity ($0.02–0.1$ g of contaminant L^{-1} day^{-1})

Limited supply of macronutrients (nitrogen, phosphorus) and micronutrients (iron, manganese, etc.)

Peat/compost

Advantages

Commercial technology

Suitable for low contaminant concentrations

Low cost media

Disadvantages

Prone to channelling and maldistribution

Limited ability to neutralize acidic degradation products

Humidification of air required to prevent drying of bed

Eventual media replacement required

Low degradation capacity ($0.02–0.4$ g L^{-1} day^{-1})

Limited supply of macronutrients (nitrogen, phosphorus) and micronutrients (iron, manganese, etc.)

Synthetic support media

Advantages

Can be easily cleaned

Fast start-up of biofilter

Can handle high contaminant concentrations (> 25 ppmv)

Cheaper than activated carbon-coated media

Can degrade contaminants requiring cometabolites by supplying it with trickling nutrients

pH can be controlled

Can be seeded with preacclimated microorganisms

Both macro- and micronutrients can be supplied continuously by trickling liquid

High degradation capacity ($0.2–0.7$ g L^{-1} day^{-1})

Disadvantages

Higher media cost than soil, peat, compost

Eventual plugging of bed unless support media is properly designed

Activated carbon coated synthetic support media

Advantages

High adsorption capacity for most contaminants

Good biomass adhesion

Fast start-up of biofilter

Can handle high contaminant concentrations (>200 ppmv) and pH can be controlled

Can degrade contaminants requiring cometabolites by preadsorption

Can be seeded with preacclimated microorganisms

Table 3.8 (continued)

Soil
Both macro- and micronutrients can be supplied continuously by trickling liquid
High degradation capacity ($0.4–1.5$ g L^{-1} day^{-1})
Disadvantages
Higher media cost than soil, peat, compost
Eventual plugging of bed requiring cleaning or media replacement unless support media is properly designed

pathogens of workers on-site and individuals off-site. This issue has been addressed by several European studies (Adler 2001). It has been observed that biofilter exhaust contains bacteria as well as fungal spores but particularly for raw gases containing high concentrations of microorganisms such as from composting and rendering operations, biofilters generally reduce the levels of entrained microorganisms. Concentrations of microorganisms in biofilter exhaust are typically only a little higher than in ambient air and significantly lower than in ambient air near composting facilities. The potential for unhealthful exposure of off-site persons to airborne microorganisms from a biofilter is low because of dispersion. But the high concentrations of microorganisms, particularly fungal spores, in filter media could expose workers during installation, monitoring, and possibly fluffing of media, since these activities tend to release some of the fungal spores into the ambient air. Therefore, the use of respiratory protection by workers involved in such activities is recommended.

Several companies and equipment manufacturers supply biofiltration services. Few manufacturing companies and some engineering and design firms have developed in-house capabilities for biofilter system testing and design (Adler 2001). Many suppliers also offer biofilter engineering and design services, but typically are limited to offering basic system design. The complexity of the application will probably determine whether engineering and design expertise is essential. For relatively common and simple applications such as off-gas treatment from a leaking underground storage tank remediation system, many vendors offer readily available off-the-shelf systems. Presently, the industry is undergoing consolidation, and some of the smaller companies with relatively weaker capabilities to provide support are disappearing but this is expected to change significantly in the USA over the next few years (Adler 2001).

Compared to other conventional physicochemical treatments biofiltration has some advantages and disadvantages that are given in Table 3.9.

3.2   Biotrickling Filter

Biotrickling filters are characterized by a continuous aqueous phase trickling throughout the reactor bed (le Reux 2010). In many cases, biotrickling filters have been shown to be more effective than conventional biofilters, and, in spite of their higher operational and initial costs, they are often preferred (Vedova 2008). The

Table 3.9 Advantages and disadvantages of biofiltration

Advantages
The pollutant is mineralized, forming principally carbon dioxide and water plus a little additional biomass
There are no secondary pollutants formed as a result of the treatment process—such as NOX from incineration—and the pollutant is not simply transferred to a different phase for further treatment—as in scrubbing or adsorption
Capital and operating costs are both modest compared with competing technologies, including biological technologies such as biotrickling filters or bioscrubbers, which are more complex and incur additional utility and maintenance expenses
Large volumes of gas can be treated economically
Lower chemical usage
Degradation of sparingly soluble pollutants or those in very low concentrations is possible
Biofiltration units can be designed to fit in shape and size to the industrial unit setting, optimizing spaces
System versatile to treat odours, toxic compounds, and VOCs efficiency > 90% for low contaminant conc. ($\ll 1000$ ppm)
Possibility of different media, microorganisms, and operational conditions for many emission points
Disadvantages
Only gas streams at moderate temperatures can be treated
Application is limited to fairly low concentrations of contaminant—higher concentrations inhibit metabolism and may injure the microorganisms
Long-term control is difficult
Recovery times after periods of nonuse or on initial start-up can be long
Long residence times and consequently large units can be required for treatment of recalcitrant compounds
System is not fitted for compounds, which have low adsorption and degradations rates, mainly chlorinated VOCs
Large biofilter units or large areas are required to treat contaminated sources with high chemical emissions
Sources of emission that vary severely or produces spikes can be detrimental to the biofilter performance and to the microbial population
Biofilters require long periods of acclimation for microbial population, weeks or even months, mainly for VOC treatment

working principles are the same compared with conventional biofilters but the presence of a trickling liquid imposes some different design conditions. Trickling liquid enhances the risk of bed compaction. Because of this reason, the packing is normally constituted by inert or synthetic material. The control of the pH, nutrients, and presence of toxics is allowed by the analysis of the trickling solution which is usually recirculated. Trickling liquid also removes toxic or acidifying by-products from inside the bed. A schematic biotrickling filter is shown in Fig. 3.3. Water phase provides the right moisture for biomass activity and no prehumidification system is required. Also, some mechanical pretreatment to remove dust, ashes, or grease is not essential since the water phase is useful to remove them from inside the bed. Gas flow can be cocurrent or countercurrent with respect to the liquid phase. There is no experimental data available indicating the best configuration. Despite of an increase in the mass transfer rate, countercurrent efficiency is affected by the presence of a

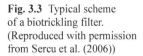

Fig. 3.3 Typical scheme
of a biotrickling filter.
(Reproduced with permission
from Sercu et al. (2006))

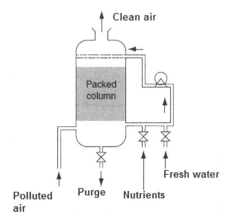

large amount of pollutants recirculated at the top of the reactor with the recycled liquid. Encountering an upward clean gas flow, the solute pollutant can be easily stripped which reduces the functioning of the bioreactor. For this reason, downward flow is generally favoured. The water pump for the recycling of the leachate is a significant aspect in the use of biotrickling filters, particularly in full-scale plants (Webster et al. 1999). Figure 3.4 shows the evolution of a biotrickling filter packed with Raschig rings, after colonization by bacteria, due to clogging and channelling.

Inert packing is usually preferred in biotrickling filters. Such carrier has good mechanical properties, low weight, and chemical stability. Furthermore, it is suitable for the biomass attachment. Clogging problem is found to be serious in biotrickling filters compared to conventional biofilters (Kennes and Veiga 2001; Schroeder 2002). Biomass growth can reduce the cross-sectional area increasing the pressure drop throughout the reactor. The most used packing materials are lava rock, plastic rings, activated carbon, ceramic rings, polyurethane foams, and perlite (Kennes and Veiga 2001, 2002). Inert packings have many disadvantages. These require biomass inoculum, have low nutrients and water retention properties, and lower specific surface area.

Biotrickling filters normally require a biomass inoculum since they are usually filled with inert carrier. Microorganisms can be obtained from other biotrickling filters, from waste water treatment plants or from a laboratory selection and culture. Start-up period is strongly correlated to biomass origin. Compared with activated sludges, pure cultures or cultures preselected with the pollutant of interest have shown shorter times to reach the maximum efficiency. After start-up period, activated sludges have shown better resistance to inlet disturbances and higher stability in long-term operations. Both fungi and bacteria have been used, but also higher microbes are generally present. Pure culture in biotrickling filters are often substituted by a mixed biomass since the system is open and contamination is possible. Fungi have shown higher removal efficiency particularly with hydrophobic pollutants (Cox and Deshusses 1998; Vargara-Fernandez and Revah 2007).

Biomass growth is a big problem during long-term operations. Since they have a large amount of nutrients at their disposal, microorganisms can grow fast, reducing

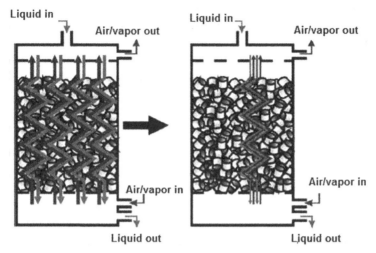

Fig. 3.4 Evolution of a biotrickling filter packed with Raschig rings, after colonization by bacteria, due to clogging and channelling. (Reproduced with permission from Sercu et al. (2006))

the cross-sectional area. Several studies have examined different systems to control biomass growth. These include chemical, biological, and physical–mechanical methods. But all these methods partially or temporarily reduce performance of biotrickling filter. All the techniques reported here have been applied only in laboratory tests, and their real efficacy on full-scale plants should be still tested. Mechanical removal methods include backwashing and periodic stirring of the carrier (Alonso et al. 1997; Cox and Deshusses 1999; Iliuta and Larachi 2004; Wubker et al. 1997; Zhu et al. 1998). These techniques are simple and effective. However, these are very expensive. Moreover, backwashing can be applied only with no-fluidizing packing. After the mechanical removal, biotrickling filters need some days to reach the elimination capacities they had before the treatment. For conventional biofilters also, backwashing has been used. Some chemicals have been used into the trickling liquid to remove biomass from the biotrickling filter. Attempts were made with sodium hydroxide solution in toluene degrading bioreactor (Weber and Hartmans 1996). Sodium hydroxide (0.1 M) was provided to the system every two weeks and 230 g of dry-weight biomass was removed. One day after the chemical wash, the elimination capacity was completely recovered. Cox and Deshusses (1999) used different mixtures of sodium hydroxide, sodium dodecylsulphate, sodium azide, hydrogen peroxide, ethanol, saturated iodine, and ammonia. Many of these attempts completely deactivated the biomass. The most promising chemical was found to be sodium hypochlorite. Kennes and Veiga (2002) reported that reducing the addition of some important nutrients, particularly nitrogen, may be a good way to reduce the growth of biomass but elimination capacity is also strongly reduced. With mixed and complex biomass, pluricellular microorganisms are also present besides fungi and bacteria. Protozoal predation is found to be an economical and environmentally friendly system for controlling the growth of biomass in biofilters and biotrickling filters (Kennes and Veiga 2002).

Several transport mechanisms have been reported to operate either simultaneously or sequentially in a biotrickling filter (Govind 2000; Govind and Narayan 2006). These mechanisms involve: diffusion of the contaminants from the bulk gas flow to the active biofilm surface, solubilization of the contaminants into the water content of the biofilms, direct adsorption of the contaminants on the surface of the support media, diffusion and biodegradation of the contaminants in the active biofilm, sorption of the contaminants directly on the biofilm surface, and surface diffusion of the contaminants in the support media surface and back diffusion of the adsorbed contaminants from the support media surface into the active biofilms. The effect of adsorption of contaminants on support medium surface, surface diffusion, and back-diffusion of the adsorbed contaminants from the support media surface into the active biofilms, mainly occurs in activated carbon-coated support media and contaminants which have affinity for the support media surface.

Biotrickling filters are found to be more effective than conventional biofilter. This is due to the recirculated trickling water which is a an effective way to:

- Control the pH inside the reactor
- Introduce nutrients and additional minerals for the biomass growth
- Remove any toxic or inhibiting substances from inside the packing

The control of the process becomes easier during long-term operation and high elimination capacity can be achieved. The right amount of water can be determined only experimentally (Kennes and Thalasso 1998). Zhu et al. (1998) have reported that biofilm drying should be avoided but an excessive water can reduce the specific area which increases the mass transfer resistance of the pollutants. Operating costs are affected by the amount of substances required for the process control. Because of the addition of alkali for maintaining the pH, the degradation of air streams containing chlorinated compounds can be much more costly if compared with other different waste gases (Deshusses and Webster 2000).

Only few studies have studied the contribution of the recycle liquid to the overall removal efficiency. Cox et al. (2000) have reported that the suspended biomass in the liquid has an average specific activity 20 times higher than that of the attached biomass. Furthermore, such biomass does not originate from a detachment of the biofilm, but it is a result of a specific growth. But its contribution to the removal efficiency is insignificant, because of the very low amount of suspended biomass.

Biotrickling filters can better face the control of some parameters within the reactor, such as pH, temperature, mineral media, and salinity compared to conventional biofilters, thanks to the moving liquid phase. It allows the washing out of intermediates and products of the cellular metabolism and the supplying of nutrient media into the system. Finally, they have shown better biomass adaptation capacity. Major problems in the use of biotrickling filters concern the degradation of the packing material. Biomass in biotrickling filters can grow faster compared to conventional biofilters. This may result in reduction of the packing specific area, producing clogging problems and the formation of anaerobic zones. This may result in decrease in the removal efficiency.

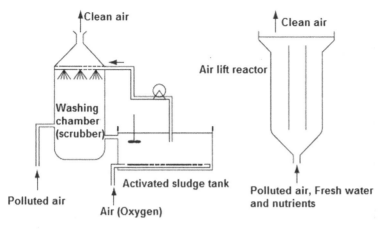

Fig. 3.5 Schematic of biowashers: to the *left*, a two-unit reactor with scrubber and activated sludge; to the *right*, a single-unit airlift reactor. (Reproduced with permission from Sercu et al. (2006))

3.3 Bioscrubbers

Bioscrubbers are also named biowashers or circulating-bed biofilm reactors. These reactors are known for their good mass transfer properties. Although the biomass in an airlift reactor is attached to the particles, they are still suspended in the circulating liquid phase. Bioscrubbers are reactors in which absorption and reaction occur in two different units and they are mainly employed with high-solubility and low-volatility pollutants (Peixoto and Ribeiro Pinto 2012; Shareefdeen et al. 2005; Sercu et al. 2006). Bioscrubbers use two separate reactors for treatment of volatile organic compounds components in off-gasses (Fig. 3.5). In the first reactor, a scrubber, contaminated gas is contacted with an aqueous solution, with or without suspended microbes, by means of a fine spray usually onto an inert packing material. This results in contaminant absorption from the gas phase to the aqueous phase. The aqueous phase is then transferred to a fixed-film bioreactor where contaminants are degraded. The water may be recycled back to the sprayer (Van Groenestijn and Hesselink 1993). Bioscrubbers have the ability to control nutrients, pH, and separate operational strategies for the two-reactor system but have the lowest gas–liquid surface area for mass transfer (Van Groenestijn and Hesselink 1993).

Depending on the characteristics of the waste gas, the performance of the absorption and reaction units can be separately increased. Mass transfer can be enhanced by suitable packaging or by increasing the number of theoretical plates. In the same way, reaction can be optimized by using a selected biomass or controlling the pH and temperature. But the requirement of high investment and operating costs with the lack of knowledge of the process have limited the diffusion of this equipment.

Bioscrubbers are found to be effective in waste water treatment plants, even for high strength odourous air streams and are a relatively new technology in the USA.

Table 3.10 Characteristics, advantages, and disadvantages of the three techniques used to treat polluted air. Based on Van Groenestijin and Hesselink (1993)

Biofiltration	Biotrickling filtration	Bioscrubbing
Characteristics	*Characteristics*	*Characteristics*
Immobolized biomass	Suspended biomass	Immobolized biomass
Immobile water phase	Mobile water phase	Mobile water phase
Single reactor	Two reactors	Two reactors
Advantages	*Advantages*	*Advantages*
Low investment and running costs	Simple and flexible design	
Operating at room temperature	pH, temperature, salinity, and mineral media control	Smaller volumes
Effective at high humidity levels	High EC for H2S	Good control of the process
Effective at low concentrations and high flow rates	Better biomass adaptation capacity	Suitable for high pollutant concentration
Safe	Washing-out of intermediates, by-products, toxics	Good stability
Generation of no-toxic byproducts		No problems concerning the carrier
		Well-established design
Disadvantages	*Disadvantages*	*Disadvantages*
Sensible at concentration and flow rates peaks	Adsorption may be the RDS of the process	Higher costs
Long start-up period	Excessive biomass growth can cause clogging	High biomass generation
Lack of knowledge	Media requires replacement	More complicated start-up procedure
Sensible at climatic changes	Pilot- and Full-Scale plant still developing	Effective for pollutants with a dimensionless Henry's coefficient < 0.01
	More expensive and complex than Conventional Biofilter	Wash-out of microorganism possible

EC Eddy correlation technique; *RDS* Rate determining step

There are several bioscrubber manufacturers in the USA. Based on the biological population, bioscrubbers fall into two categories. Autotrophic organisms eliminate hydrogen sulphide and other inorganic compounds and oxidize sulphides to elemental sulphur or sulphate. Heterotrophic organisms eliminate volatile organic compounds and have little effect on hydrogen sulphide. The two systems may be combined into a two-stage system, where treatment of hydrogen sulphide and volatile organic compounds is required. The systems use intermittent irrigation, with the biologically active solution trickling over the medium bed to keep the biomass wet. This is done to promote sloughing of the biomass and to supply fresh nutrients to the biomass. Bioscrubbers are only occasionally used, particularly for removing high concentrations of highly water-soluble compounds. They have been used in the treatment of waste gases from incinerators and foundry industry (amines, phenol, formaldehyde, and ammonia).

The advantages and disadvantages of the three techniques used to treat polluted air are presented in Table 3.10.

References

Adler SF (2001) Biofiltration a primer. Chem Eng Prog 97(4):33–41

Alonso C, Suidan MT, Sorial GA, Smith FL, Biswas P, Smith PJ, Brenner RC (1997) Gas treatment in trickle-bed biofilters: biomass, how much is enough? Biotechnol Bioeng 54:583–594

Andreoni V, Origgi G, Columbo M, Calcaterra E, Columbi A (1997) Characterization of a biofilter treating toluene contaminated air. Biodegradation 7:397–404

Anit S, Artuz R (2000) Biofiltration of air. Rensselaer Institute. www.rensselaer.edu/dept/chemeng/Biotech-Environ/MIS/biofilt/biofiltration.htm. Accessed 12 March 2013

Anon (1991) Air pollution control may be reduced with biotechnology. RMT Netw 6(1):5–8

Arulneyam D, Swaminathan T (2003) Biodegradation of methanol vapor in a biofilter. J Environ Sci 15:691–696

Arulneyam D, Swaminathan T (2005) Biodegradation of mixture of VOC's in a biofilter. J Environ Sci 16:30–33

Bajpai P, Bajpai PK, Kondo R (1999) Biotechnology for environmental protection in the pulp and paper industry. Springer, Germany

Barnes JM, Apel WA, Barrett KB (1995) Removal of nitrogen oxides from gas streams using biofiltration. J Hazard Mat 41:315–326

Barshter DW, Paff SW, King AB (1993) Biofiltration: room temperature incineration. In: Proceedings 86th annual meeting of the air & waste manage. Assn., Denver, Colorado

Becker M, Rabe R (1997) Emission of fungal spores from a biofilter. In: Prins WL, van Ham J (eds) Biological waste gas cleaning, proceedings of an international symposium. VDI, Düsseldorf, pp 221–224

Bendinger B, Kroppenstedt RM, Klatte S, Altendorf K (1992) Chemotaxonomic differentiation of coryneform bacteria isolated from biofilters. Int J Syst Bacteriol 42(3):474–486

Bohn HL (1975) Soil and compost filters for malodorant gases. J Air Waste Manag Assoc 25:953–955

Bohn HL (1992) Considering biofiltration for decontaminating gases. Chem Eng Prog 88:34–40

Bohn HL (1993) Biofiltration: design principles and pitfalls. In: Proceedings of 86th Annual Meeting of Air and Waste Manage. Assoc. Denver, Paper #93-TP- 52 A.01

Bohn HL (1996) Biofilter media. In: Proceedings of the 89th Annual Meeting and Exhibition of Air and Waste Management Association, Nashville, US

Bohn H, Bohn R (1988) Soil beds weed out air pollutants. Chem Eng 95(4):73–76

Brauer H (1986) Biological purification of waste gases. Int Chem Eng 26(3):387–395

Burgess JE, Parsons SA, Stuetz RM (2001) Developments in odour control and waste gas treatment biotechnology: a review. Biotechnol Adv 19(1):35–63

Cáceres M, Morales M, San MR, Urrutia H, Aroca G (2010) Oxidation of volatile reduced sulphur compounds in biotrickling filter inoculated with Thiobacillus thioparus. Electron J Biotechnol 13(5). http://dx.doi.org/10.2225/vol13-issue5-fulltext-9

Campbell HJ, Connor MA (1997) Practical experience with an industrial biofilter treating solvent vapor loads of varying magnitude and composition. Pure Appl Chem 69(11):2411–2424

Carlson DA, Leiser CP (1966) Soil bed for control of sewage odors. J Wat Poll Contr Fed 38:829–833

Chitwood DE (1999) Two-stage biofiltration for treatment of POTW off-gases. Thesis, University of Southern California, Los Angeles

Cho KS, Hirai M, Shoda M (1991) Removal characteristics of hydrogen sulphide and methanethiol by Thiobacillus sp. isolated from peat in biological deodorization. J Fermen Bioeng 71:44–49

Chung YC (2007) Evaluation of gas removal and bacterial community diversity in a biofilter developed to treat composting exhaust gases. J Hazard Mater 144:377–385

Chung YC, Huang C, Tseng CP (2001) Biological elimination of H2S and NH3 from waste gases by biofilter packed with immobilized heterotrophic bacteria. Chemosphere 43:1043–1050

Cloirec PL, Humeau P, Ramirez-Lopez EM (2001) Biotreatments of odours and performances of a biofilter and a bioscrubber. Water Sci Tech 44(9):219–226

Cohen Y (2001) Biofiltration-the treatment of fluids by microorganisms immobilized into the filter bedding material: a review. Bioresour Technol 77(3):257–274

Cox HHJ, Deshusses MA (1998) Biological waste air treatment in biotrickling filters. Curr Opin Biotechnol 9:256–262

Cox HHJ, Moerman RE, van Baalen S, van Heiningen WNM, Doddema HJ, Harder W (1997) Performance of a styrene degrading biofilter containing the yeast Exophiala jeanselmei. Biotechnol Bioeng 53:259–266

Cox HHJ, Deshusses MA (1999) Chemical removal of biomass from waste air biotrickling filters: screening of chemicals of potential interest. Water Res 33(10):2383–2391

Cox HHJ, Nguyen TT, Deshusses MA (2000) Toluene degradation in the recycle liquid of biotrickling fiters for air pollution control. Appl Microbiol Biotechnol 54:133–137

Dawson DS (1993) Biological treatment of gaseous emissions. Wat Env Res 65(4):368–371

DeBont JAM, vanDijken JP, Harder W (1981) Dimethyl sulfoxide and dimethyl sulfide as a carbon, sulfur and energy source for growth of Hyphomicrobium S. J Gen Microbiol 127:315–323

Deutsch S (2006) Biofilter system lowers manure load, Odor by 90%. http://nationalhogfarmer.com/mag/farming_biofilter_system_lowers/. Accessed 20 Jan 2013

Deshusses MA (1997) Biological waste air treatment in biofilters. Curr Opin Biotechnol 8:335–339

Deshusses MA, Hammer G (1993) The removal of volatile ketone mixtures from air in biofilters. Bioprocess Eng 9:141–146

Deshusses MA, Webster TS (2000) Construction and economics of a pilot/full-scale biological trickling filter reactor for the removal of volatile organic compounds from polluted air. J Air Waste Manag Assoc 50:1947–1956

Devinny JS, Deshusses MA, Webster TS (1999) Biofiltration for air pollution control. CRC/Lewis Publisher. ISBN:1-56670-289-5

Diks RMM, Ottengraf SPP (1994) The influence of NaCI on the degradation rate of dichloromethane by Hyphomicrobium sp. Biodegradation 5:129–141

Eisenring R (1997) Technical fabrics as novel carrier materials for biofilters and biological trickling bed reactors. Wasser Luft und Boden 41(9):57–61

Ergas S, Schroeder ED, Chang DP, Scow KM (1994) Spatial distribution of microbial populations in biofilters, 94-RP115B01. In: Proceedings of the 87th annual meeting of the air & waste management association. Cincinnati, OH

Ergas SJ, Schroeder ED, Change DPY, Morton RL (1995) Control of volatile organic compound emissions using a compost biofilter. Water Environ Res 67(5):816–821

Farmer RW (1994) Biofiltration: process variables and optimization studies. M.S. Thesis, University of Minnesota. December 1994

Finn L, Spencer R (1997) Managing biofilters for consistent odor and VOC treatment. Biocycle 01:1997

Fouhy K (1992) Cleaning waste gas, naturally. Chem Eng 99(12):41–46

Furusawa N, Togashi I, Hirai M, Shoda M, Kubota H (1984) Removal of hydrogen sulfide by a biofilter with fibrous peat. J Ferment Technol 62(6):589–594

Garcia-Pena EI, Hernandez S, Favela-Torres E, Auria R, Revah S (2001) Toluene biofiltration by the fungus Scedosporium apiospermum TB1. Biotechnol Bioeng 76:61–69

Gerrard AM, Metris AV, Paca J (2000) Economic designs and operation of biofilters. Eng Economist 45:259–270

Goldstein N (1996) Odor control experiences: lessons from the biofilter. Bio Cycle 37(4):70

Goldstein N (1999) Longer life for biofilters. Bio Cycle 40(7):62

Govind R (2000) Biofiltration: an innovative technology for the future. In: Proceedings of the water environment federation, odors and VOC emissions, Water Environment Federation. pp 864–88. www.prdtechinc.com/PDF/PrdbiofilterR&DMagazinePaper.pdf. Accessed 22 Dec 2013

Govind R, Bishop DF (1996) Overview of air biofiltration—basic technology, economics and integration with other control technologies for effective treatment of air toxics. Emerging solutions VOC air toxics control. In: Proceedings Spec Conf. Pittsburgh, Pa, pp 324–350

Govind R, Narayan S (2006) Selection of bioreactor media for odor control. www.prdtechinc.com/ PDF/PRDBIOMEDIADESIGNCHAPTER.pdf. Accessed 20 Feb 2013

Hirai M, Ohtake M, Soda M (1990) Removal of kinetics of hydrogen sulfide, methanethiol and dimethyl sulfide by peat biofilters. J Ferment Bioeng 70:334–339

Hodge DS, Devinny JS (1994) Biofilter treatment of ethanol vapors. Environm Progr 13(3):167–173

Hodge DS, Medina VF, Islander R, Devinny JS (1991a) Biofiltration of Hydrocarbon Fuel Vapors in Biofiltration. Environ Technol 12:665–662

Hodge DS, Medina VF, Islander RL, Devinny JS (1991b) Treatment of hydrocarbon fuel vapors in biofilters. Environ Technol 12:655–662

Hodge DS, Medina VF, Wang Y, Devinny JS (1992). Biofiltration: application for VOC emission control. In: Wukasch RF (ed) Proceedings of the 47th industrial waste conference. Perdue University, West Lafayette, pp 609–619

Holusha J (March 1991) Using bacteria to control pollution. The New York Times 13:C6

Hort C, Gracy S, Platel V, Moynault L (2009) Evaluation of sewage sludge and yard waste compost as a biofilter media for the removal of ammonia and volatile organic sulfur compounds (VOSCs). Chem Eng J 152:44–53

Irvine RL, Moe WM (2001) Periodic operation for enhanced performance during unsteady-state loading conditions. Water Sci Technol 43(3):231–239

le Reux LD, Johnson ME (2010) Performance of high-rate biotrickling filter under ultra-high loadings at a municipal WWTP, odors and air pollutants 2010, water environment federation

Iliuta I, Larachi F (2004) Transient biofilter aerodynamics and clogging for VOC degradation. Chem Eng Sci 59:3293–3302

Janni KA, Nicolai RE, Jacobson LD, Schmidt DR (1998) Low cost biofilters for odor control in Minnesota. Final Report 14 August 1998. St. Paul, Minnesota: Biosystems and Agricultural Engineering Department, University of Minnesota

Janni KA, Maier WJ, Kuehn TH, Yang CH, Bridges BB, Vesley D (2001) Evaluation of biofiltration of air, an innovation air pollution control technology. ASHRAE Trans 107(1):198–214

Kanagawa T, Kelly DP (1986) Breakdown of dimethyl sulfide by mixed cultures and by Thiobacillus thioparus. FEMS Microbiol Lett 34:13–19

Kanagawa T, Mikami E (1989) Removal of methanethiol, dimethyl sulfide, dimethyl disulfide and hydrogen sulfide from contaminated air by Thiobacillus thioparus TK-m. Appl Environ Microbiol 55(3):555–558

Kennes C, Thalasso F (1998) Waste gas biotreatment technology. J Chem Technol Biotechnol 72:303–319

Kennes C, Veiga MC (2001) Bioreactors for waste gas treatment. Kluwer Acadamic, Dordrecht

Kennes C, Veiga MC (2002) Inert filter media for the biofiltration of waste gases—characteristics and biomass control. Rev Environ Sci Biotechnol 1:201–214

Kennes C, Veiga MC, Yaomin J (2007) Co-treatment of hydrogen sulfide and methanol in a single-stage biotrickling filter under acidic conditions. Chemosphere 68(6):1186–1193

Kim H, Kim YJ, Chung JS (2002) Long term operation of a biofilter for simultaneous removal of H2S and NH3. J Air Waste Manag Assoc 52(12):1389–1398

Kiared K, Bibeau L, Brzenzinski R, Viel G, Heitz M (1996) Biological elimination of VOCs in biofilter. Environ Prog 15(3):148–152

Kiared K, Wu G, Beerli M, Rothenbuhler M, Heitz M (1997) Application of biofiltration to the control of VOC emissions. Environ Technol 18(1):55

Kirchner K, Hauk G, Rehm HJ (1987) Exhaust gas purification using immobilized monocultures (biocatalyst). App Microbiol Biotechnol 26:579–587

Kraakman NJR (2004) H2S and odor control at wastewater collection systems: an on-site study on the robustness of a biological treatment. Poster presented at the 2004 USC-CSC-TRG Conference on biofiltration, Santa Monica, USA, October 20–22

Kraakman NJR (2005) Biotrickling and bioscrubbers applications to control odor and air pollutants: developments, implementation issues and case studies. In: Shareefdeen Z, Singh A (eds) Biotechnology for odour and air pollution control. Springer, Heidelberg, pp 355–379

Krishna M, Philip L, Venkobachar C (2000) Performance evaluation of a thiobacillus denitrifiers immobilized biofilter for the removal of oxides of nitrogen. Presented at the USC-TRG Conference on biofiltration (an air pollution control technology), pp 191–199

Lehtomaki J, Torronen M, Laukkarinen A (1992) A feasibility study of biological waste air purification in a cold climate. In: Dragt AJ, van Ham J (eds) Biotechniques for air pollution abatement and odour control policies. Elsevier Science Publisher B. V., Amsterdam, pp 131–134

Leson G, Winer AM (1991) Biofiltration: an innovative air pollution control technology for VOC emissions. J Air Waste Manag 41:1045–1054

Lipski A, Klatte S, Bendinger B, Altendorf K (1992) Differentiation of Gram negative, nonfermentative bacteria isolated from biofilters on the basis of fatty acid composition, quinone system, and physiological reaction profiles. Appl Environ Microbiol 58(6):2053–2065

Lee SK, Shoda M (1989) Biological deodorization using activated carbon fabric as a carrier of microorganisms. J Ferment Bioeng 68(6):437–442

Luo J, Lindsey S (2006) The use of pine bark and natural zeolite as biofitlter media to remove animal rendering process odours. Bioresour Tech 97(13):1461–1469

Luo J, Oostrom A (1997) Biofilters for controlling animal rendering odour: a pilot scale study. Pure Appl Chem 69(11):2403–2410

Mallakin A, Ward OP (1996) Biodegradation of gasoline and BTEX in a microaerophilic biobarrier. J Biodegrad 10(5):341–352. (Cover Date)

Marsh A (1994) Biofiltration for emission abatement. Eur Coat J 7(8):528–536

McNevin D, Barford J (2000) Biofiltration as an odour abatement strategy. Biochem Eng J 5(3):231–242

Miller GY, Maghirang RG, Riskowski GL, Heber AJ, Robert MJ, Muyot MET (2004) Influences on air quality and odor from mechanically ventilated swine finishing buildings in Illinois. Food Agric Environ 2(2):353–360

Morgenroth E, Schroeder ED, Chang DPY, Scow KM (1996) Nutrient limitation in a compost biofilter degrading hexane. J Air Waste Manag Assn 46:300–308

Moreno L, Predicala B, Nemati M (2010) Laboratory, semi-pilot and room scale study of nitrite and molybdate mediated control of H2S emission from swine manure. Biores Technol 101:2141–2151

Morton RL, Caballero RC (1997) Removing hydrogen sulfide from wastewater treatment facilities air process streams with a biotrickling filter. 90th annual meeting, Air and waste process streams with a biotrickling filter, 90th annual meeting, Air and waste management association

Mueller JC (1998) Biofiltration of gases-A mature technology for control of a wide range of air pollutants, report to the National Research Council of Canada and British Columbia Ministry of Advanced Education and Job Training, Project Number: 2-51-797

Nanda S, Sarangi PK, Abraham J (2012) Microbial biofiltration technology for odour abatement: an introductory review. J Soil Sci Environ Manag 3(2):28–35. http://www.academicjournals. org/JSSEM. doi:10.5897/JSSEM11.090. ISSN 2141-2391. Accessed 22 Dec 2013

Naylor LM, Kuter GA, Gormsen PJ (1988) Biofilters for odor control: the scientific basis. In compost facts. Glastonbury, Conn.: International Process System, Inc

Norman CW (2002) The effect of periodic operation on biofilters for removal of methyl ethyl ketone from contaminated air. Thesis Master of Science Louisiana State University. etd.lsu.edu/ docs/available/etd-0418102141608/.../Norman_thesis.pdf. Accessed 22 Dec 2013

Ottengraf SPP (1986) Exhaust gas purification. In: Schonborn W (vol ed), Rehm H-J, Reed G (series eds). Biotechnology Vol. 8 (Microbial Degradations). VCH, Weinheim, pp 425–452. (Chap. 12)

Ottengraf SPP (1987) Biological system for waste gas elimination. TIBTECH 5:132–136

Ottengraf SPP, Konongs JHG (1991) Emission of microorganisms from biofilters. Bioprocess Eng 7:89–96

Ottengraph SPP, Van Denoever AHC (1983) Kinetics of organic compound removal from waste gases with a biological filter. Biotechnol Bioeng 25:3089–3102

Ottengraf SPP, Meesters JPP, van den Oever AHC, Rozema HR (1986) Biological elimination of volatile xenobiotic compounds in biofilters. Bioprocess Eng 1:61–69

Peixoto JM, Ribeiro Pinto JC (2012) Biofilm growth and hydrodynamic behaviour in the biological plate tower (BPT) with and without hanging biomass (BPT-HB).Water Sci Technol 66(8):1678–1683

Pomeroy D (1982) Biological treatment of odorous air. J Water Pollut Control Fed 54:1541–1545

Pond RL (1999) Biofiltration to reduce VOC and HAP emissions in the board industry. Tappi J 82(8):137–140

Rafson HJ (1998) Odor and VOC control handbook. McGraw-Hill, New York

Rappert S, Muller R (2005) Microbial degradation of selected odorous substances. Waste Manag 25:940–954

Reichert K, Lipski A, Altendorf K (1997) Degradation of dimethyl disulfide and dimethyl sulfide by Pseudonocardia strains. In: Biologische Abgasreinigung. VDI-, Düsseldorf, pp 269–272

Rogers GL, Klemetson SL (1985) Ammonia removal in selected aquaculture water reuse biofilters. Aqua Eng 4(2):135–154

Rozich A (1995) Tackle airborne organic vapors with biofiltration. Environ Eng World 1:32–34

Saravanan V, Rajasimman M, Rajamohan N (2010) Biofiltration kinetics of ethyl acetate and xylene using sugarcane bagasse based biofilter. Chem Eng Res Bull 14:51–57

Schroeder ED (2002) Trends in application of gas-phase bioreactors., Rev Environ Sci Biotechnol 1:65–74

Sercu B, Peixoto J, Demeestere K, van Helst T, van Langenhove H (2006) Odors treatment: biological technologies. In: Nicolay X (ed) Odors in the food industry. Springer, USA pp 125–158

Shareefdeen Z, Baltzis BC, Oh YS, Bartha R (1993) Biofiltration of methanol vapor. Biotechnol Bioeng 41:512–524

Shareefdeen Z, Herner B, Singh A (2005) Biotechnology for air pollution control—an overview. In: Shareefdeen Z, Singh A (eds) Biotechnology for odor and air pollution control. Springer, Berlin

Singhal V, Singla R, Walia AS, Jain SC (1996) Biofiltration—an innovative air pollution control technology for H2 S emissions. Chem Eng World 31(9):117–124

Sivela S, Sundman V (1975) Demonstration of Thiobacillus type bacteria which utilize methyl sulfides. Arch Microbiol 103:303–304

Smet E, Van Langenhove H, De Bo I (1999) The emission of volatile compounds during theaerobic and the combined anaerobic/aerobic composting of biowaste. Atmospher Environ 33:1295–1303

Smith NA, Kelly DP (1988a) Isolation and physiological characterization of autotrophic sulfur bacteria oxidizing dimethyl disulfide as sole source of energy. J Gen Microbiol 134:1407–1417

Smith NA, Kelly DP (1988b) Mechanism of oxidation of dimethyl disulfide by Thiobacillus thioparus strain E6. J Gen Microbiol 134:3031–3039

Soccol CR, Woiciechowskil AL, Vandenberghel LPS, Soares M, Neto GN, Thomaz Soccol V (2003) Biofiltration: an emerging technology. Indian J Biotechnol 2(3):396–410

Song J, Kinney KA (2002) A model to predict long-term performance of vapor-phase bioreactors: a cellular automaton approach. Environ Sci Technol 36(11):2498–2507

Sorial G, Smith F, Suidan M (1997) Performance of a peat biofilter: Impact of the empty bed residence time, temperature, and toluene loading. J Hazardous Mater 53(1–3):19–33

Spigno G, Pagella C, Fumi MD, Molteni R, De Faveri DM (2003) VOCs removal from waste gases: gas-phase bioreactor for the abatement of hexane by Aspergillus niger. Chem Eng Sci 58:739–746

Suylen GM, Stefess GC, Kuenen JG (1986) Chemolithotropic potential of a Hyphomicrobium species capable of growth on methylated sulfur compounds. Arch Microbiol 146:192–198

Suylen GM, Large PJ, vanDijken JP, Kuenen JG (1987) Methyl mercaptan oxidase, a key enzyme in the metabolism of methylated sulfur compounds by Hyphomicrobium EG. J Gen Microbiol 133:2989–2997

Swanson W, Loehr R (1997) Biofiltration: Fundamentals, design and operation principles, and applications. J Environ Eng ASCE 123:538–546

Thorsvold BR (2011) Biological odor control systems: a review of current and emerging technologies and their applicability nfo.ncsafewater.org/SharedDocuments/Web Site Documents/Annual Conference/AC 2011 Papers/WW_T.pm_03.45_Thorsvold.pdf. Accessed 20 March 2014

Tolvanen OK, Hanninen KI, Veijanen A, Villberg K (1998) Occupational hygiene in biowaste composting. Waste Manag Res 1994(16):525–540

Van Groenestijn JW, Hesselink Paul GM (1993) Biotechniques for air pollution control. Biodegradation 4:283–301

Van Lith C, Leson G, Michelson R (1997) Evaluating design options for biofilters. J Air Waste Manag Assn 47:37–48

Van den Berg L, Kennedy KJ (1981) Support materials for stationary fixed films reactors for high ratemethonogen fermentations. Biotech Lett 3:165–170

Van Langenhove H, Wuyts E, Schamp N (1986) Elimination of hydrogen sulfide from odorous air by a wood bark biofilter. Water Res 20:1471–1476

Vargara-Fernandez A, Revah S (2007) Modeling of fungal biofilter for the abatement of hydrophobic VOCs. In: Proceedings of the 2nd international congress biotechniques for air pollution control, A Coru˜na, Spain, October 3–5

Vedova LD (2008) Biofiltration of industrial waste gases in trickle-bed bioreactors, Case study: trichloroethylene removal. Universit`a degli Studi di Padova sede amministrativa: Universit`a degli Studi di Padova sede consorziata: Universit`a degli Studi di Trieste DICAMP: Dipartimento di Ingegneria Chimica, dell'Ambiente e delle Materie Prime. Scuola di Dottorato in Ingegneria Industriale Indirizzo Ingegneria Chimica XX Ciclo

Wani A, Branion R, Lau AK (1997) Biofiltration: A promising and cost-effective control technology for odors, VOCs and air toxics. J Environ Sci Health 32(7):2027–2055

Weber FJ, Hartmans S (1996) Prevention of clogging in a biological trickle-bed reactor removing toluene from contaminated air. Biotechnol Bioeng 50:91–97

Webster TS, Cox HHJ, Deshusses MA (1999) Resolving operational and performance problems encountered in the use of a pilot/full-scal biotrickling filter reactor. Environ Prog 18(3):162–172

Weigner P, Paca J, Loskot P, Koutsky B, Sobotka M (2001) The start-up period of styrene degrading biofilters. Folia Microbiol 46(3):211–216

Woertz J, Kinney K (2000) Use of the fungus j & cop/i/a/a lecanii-comi to degrade a mixture of VOCs. Presented at the USC-TRG conference on biofiltration (an air pollution control technology), pp 151–158

Williams TQ, Miller FC (1992) Odor control using biofilters. Bio Cycle 33(10):72–77

Wright P, Bustard M, Meeyoo V (2000) Environmental pollution abatement technologies for the new millennium: novel vapor phase biofiltration of VOCs and other waste gases. In: Proceedings of the air and waste management association conference and exhibition, pp 2157–2168

Wubker SM, Laurenzis A, Werner U, Friedrich C (1997) Controlled biomass formation and kinetics of toluene degradation in a bioscrubber and in a reactor with a periodically moved tricklebed. Biotechnol Bioeng 55:686–692

Xie B, Liang SB, Tang Y, Mi WX, Xu Y (2009) Petrochemical wastewater odor treatment by biofiltration. Biores Technol 100:2204–2209

Yang Y, Allen ER (1994) Biofiltration control of hydrogen sulphide design and operation parameters. J Air Waste Manag Assoc 44:863–868

Zilli M, Fabiano B, Ferraiolo A, Converti A (1996) Macro-kinetic investigation on phenol uptake from air by biofiltration: influence of superficial gas flow rate and inlet pollutant concentration. Biotechnol Bioeng 49(4):391–398

Zhu X, Alonso C, Suidan MT, Cao H, Kim BJ, Kim BR (1998) The effect of liquid phase on VOC removal in trickle-bed biofilters. Water Sci Technol 38(3):315–322

Chapter 4
New Reactors

Besides the three main bioreactor configurations for air pollution control, many other alternatives have been developed which are described below (Peixoto and Ribeiro Pinto 2012; Sercu et al. 2006; Vedova 2008; De Bo 2000, 2001, 2002).

4.1 Biological Plate Tower (BPT)

Deodorization and volatile organic compounds abatement from polluted air streams can be accomplished with the biological plate tower, which has proved to be a reliable alternative to biofilters and biotrickling filters. Unlike those that are mostly applied, the biological plate tower is a nonclogging device, with constant active surface and steady performance, making it ideal for scale-up and modelling. Essentially, the biological plate tower is a pile of parallel circular plates having a single hole on the border (Peixoto and Ribeiro Pinto 2012; Sercu et al. 2006; Shareefdeen et al. 2005). The plates are placed in such a manner that the holes will alternate (180°) from one to the next plate. In this way, a cascade of liquid will go downward, changing direction from plate to plate. The gaseous stream follows the opposite direction, upward. The bacteria are attached to their top surface. Figure 4.1 shows the schematic of the flows and biofilm growth on the plates. The reactor consists of four modules of about 28.8 dm^3 each biological plate tower with 20 plates in each module. An individual plate surface area (top face) is about 40,195 mm^2 (Sercu et al. 2006). The surfaces of the plates are scratched in order to make the bacterial adhesion easier. Only two or three of the four modules are operated continuously. The other modules are kept free and ready to replace any one that reaches saturation with biomass. The operation can be kept going nearly forever in this way. The functioning is found to be fairly stable (the biofilm activity which is surface-dependent is kept approximately constant). The constant surface contact area makes it easy to model and scale-up the process. The total surface area and the space between plates can be designed for the desired operating time. In theory, the available surface in a biological plate tower is a tenth of the surface in a biotrickling filter, considering the same total volume. With the new design, a stable operation for longer periods

P. Bajpai, *Biological Odour Treatment,* SpringerBriefs in Environmental Science, DOI 10.1007/978-3-319-07539-6_4, © The Author(s) 2014

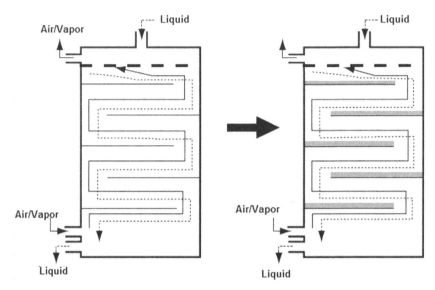

Fig. 4.1 Simplified schematics of the biological plate tower, with only five plates, to better visualize the directions of both flows and the attached biofilm on the upper surface of the plates. (Reproduced with permission from Sercu et al. 2006)

was noticed. Also, the removal of volatile organic compounds was found to be high. About 92 % removal was achieved for inlet toluene concentration of 10 g m^{-3} and empty bed residence time of 108 s. It has very good hydrodynamic performance and operates continuously without any problem. In the long term, the short area is compensated by the steady operation. Compared to the biotrickling filter, the disposal of the newly formed biomass is also much easier in this case. Unlike biofilters whose packing has to be rejected after a certain time of operation, biological plate tower biomass is withdrawn as a water-rich solid phase—the biofilm attached to the plates—and is quite easy to handle. When the thickness of the biofilm reaches the maximum value, the set of plates is simply replaced by a clean one and the biomass is dealt with outside the reactor. Sampling the biofilm for analysis is very easy. It does not oblige the operation to be stopped, or severely shaken as it happens with biotrickling filters, and any plate can be sampled. Even operation demands a constant surface of biofilm. Oxygen uptake rate measurements were done to find out if there were great activity differences between different plates and between the surface and inside the biofilm. The respiratory activity was about 0.11 mg g^{-1} s^{-1}, mass of oxygen per mass of volatile solids per time. It was identical for the superficial samples of all plates, showing some difference up to about 20 % for the lower ones where it was higher. The middle samples had almost zero activity (0.01 mg g^{-1} s^{-1} or less) and none of the base samples showed any activity. For the respirometry tests, the carbon source used was phenol. The plates at the bottom of each module had thicker biofilms compared to the upper ones, because of the higher concentration of the carbon source and oxygen in the entrance. The first module, which receives the higher dose, is the one that shows the thickest films, reaching over 15 mm until

Fig. 4.2 Photograph of the bottom plates of the first module showing the biofilm growth on the biological plate tower plates. The huge biofilm does not endanger the permeability of the system. (Reproduced with permission from Sercu et al. 2006)

needing to be replaced (Fig. 4.2). Biological plate tower solves the problem of clogging and channelling and performs very well. Mass transfer, air to liquid, is also very good due to the existence of both parallel and cross flows (Sercu et al. 2006; Mota and Peixoto 2008). The Biological plate tower showed high odour and volatile organic compounds removal. Above 90 % removal for inlet ammonium concentrations between 7.3 and 136.6 µg L^{-1} and up to 25 mg L^{-1} for toluene was observed.

Peixoto and Ribeiro Pinto (2012) enhanced the performance of BPT by changing its geometry from cylindrical (circular plates) to a rectangular cuboid (rectangular plates) and testing the hydrodynamic behaviour of cocurrent versus countercurrent flows (flooding, holdup and pressure drop) with diminished distance between adjacent plates. The diminished distance between plates was well tolerated in concurrent flow, allowing much higher quantities of biomass in the same reactor volume. With 18 and 14 mm spacing between adjacent plates, the BPT, with and without holes, was tested by Peixoto and Ribeiro Pinto (2012) for flooding, holdup, and pressure drop. Several gas and liquid flows were tested, both in cocurrent and countercurrent flows. In hydrodynamic terms, the BPT-HB with cocurrent flow was clearly the best option. Higher stability with higher flow rates and lower pressure drops were observed. The inoculum was obtained from wastewater plant activated sludge (petrochemical industry).

The research on the BPT is in progress. Studies are conducted on the following aspects:

• Assays to quantify volatile organic compounds and odour removals.
• Bacterial growth, different bacteria, plate shapes, and distances between plates.
• The bacterial growth hanging from holes in the plates and the possibility of using different bacteria in different modules.

4.2 Membrane Bioreactor

Membrane technology has been applied in waste air treatment. Membrane bioreactors are particularly effective with low soluble pollutants and the risk of clogging is completely avoided. Moreover, they are suitable to treat pollutants which require cometabolism. In membrane bioreactors, the gaseous pollutants are transferred from the gas to the liquid phase where they are degraded via a membrane (Kennes and Veiga 2001; Kraakman et al. 2007). The following categories of membrane materials have been reported for treating contaminants:

- Hydrophobic microporous membranes: Hydrophobic microporous membranes consist of a polymer matrix, for example polypropylene, polysulfone, or Teflon®, and contain pores with a diameter in the range of 0.01–1 µm. Since the membrane material is hydrophobic, the pores are filled with gas. Water does not enter the pores unless a critical pressure at the liquid side is exceeded (Reij et al. 1998). Microporous material is generally made into hollow fibres, although spiral wound and plate and frame modules have also been used (Reij et al. 1998).
- Dense phase membranes: Dense material is available as tubes with a wall thickness of at least several hundred micrometres (Reij et al. 1998).
- Composite membranes: Composite membranes are a combination of the two types (dense and microporous) and consist of a thin, selective top layer (1–30 µm) of dense material, supported by a highly porous support layer (e.g. nonwoven polyester or a microfiltration membrane (Reij et al. 1998).

Two types of biomass may be used: fixed film cultures (biofilms) and suspended growth cultures (Burgess et al. 2001; Kraakman et al. 2007). Mass transfer and kinetics of a contaminant within the membrane bioreactor module can be described as a sequence of events:

- Bulk mixing of the contaminant in the air entering the bioreactor
- Boundary layer transport
- Sorption and diffusion into the membrane
- Exit from the membrane and dissolution and diffusion into the biofilm
- Diffusion through and degradation within the biofilm
- Boundary layer transport into the liquid phase
- Subsequent mixing and degradation within the suspension

In the membrane bioreactor concept, one side of the membranes is dry. It acts as a surface for uptake of pollutants from the air flowing along the membranes. The other side of the membrane is kept wet and covered by a biofilm. Figure 4.3 shows a flat membrane bioreactor with a composite membrane. Also other configurations like hollow fibre membranes modules can be applied. Pollutants diffuse through the membrane and are consequently degraded by the *microorganisms* in the biofilm or in the recirculating aqueous phase. The microbial degradation process can be easily controlled by continuous recirculation of the aqueous phase.

Fig. 4.3 Scheme of a flat membrane bioreactor for waste gas treatment. (Reproduced with permission from Sercu et al. 2006, adaptation of De Bo 2002)

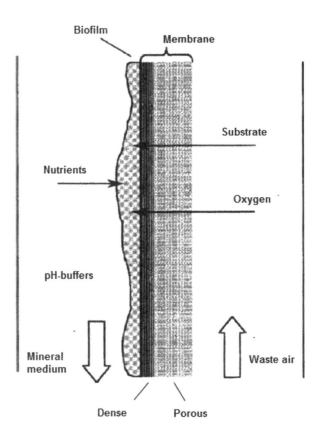

The main advantages of membrane bioreactors for waste gas treatment include (De Bo 2002):

- High specific surface area
- Ability to prevent clogging
- Good reactor control
- Physical separation of gas and biofilm
- Low pressure drop
- Absence of channelling
- Independent control of gas and liquid phase (De Bo 2002).

Potential disadvantages are:

- High investment costs
- Additional mass transfer resistance caused by the membrane
- Decreased biofilm activity as the biofilm ages
- Clumping of hollow fibre membranes at high biofilm growth

The reactor concept has potential to eliminate volatile organic compounds characterized by poor water solubility, by lack of biodegradability and toxicity

(Reij et al. 1998). A flat membrane reactor was developed and used for the degradation of dimethyl sulphide and toluene as single compounds (De Bo et al. 2002, 2003). In this case a composite membrane was used. The reactor performance was found to be stable as the clogging of the porous membrane was prevented. For dimethyl sulphide, an ECmax of 4.8 kg m^{-3} day^{-1} was obtained. This was higher than any reported figure for biofilters or biotrickling filters.

4.3   Sparged Gas Reactors

In sparged gas reactors volatile organic compounds contaminated air is passed through an aerator submerged in an aqueous-phase bioreactor (Norman 2002). This results in mass transfer from the gas phase to the aqueous phase where a suspended microbial population causes the degradation of the contaminants. Nutrient concentrations, biomass wasting, and hydraulic retention times in the reactor are controlled in the same way as the activated sludge processes used for wastewater treatment (Bielefeldt and Stensel 1999). A disadvantage of this process is that contaminated gases must be compressed, and the head loss is comparatively high.

4.4   Other Bioreactors

Suspended-growth reactors can be used to remove gaseous pollutants. The contaminated gas stream is flushed throughout a liquid phase with an active suspended biomass. Clogging and drying problems are avoided. No anaerobic regions are found to be present inside the reactor. The removal of toxins from the liquid phase and the treatment of poor soluble compounds are still problems (Neal and Loehr 2000).

Foamed emulsion biological reactors (FEBRs) use a biological foam to increase the surface area for the mass transfer. These reactors have been tested successfully with BTXs (mixtures of benzene, toluene, p-xylene, and styrene) and high removal efficiencies are also obtained with trichloroethylene (Kan and Deshusses 2003, 2006; Phipps 1998; Song and Shin 2007). Bed clogging and biomass drying are avoided. However, the requirement of the specific surfactant to generate the emulsion strongly affects the operational costs. Foam stability problem concerns for the full-scale application of this bioreactor.

Monolith bioreactors have been studied for treating toluene and methanol (Jin et al. 2007). They involve a ceramic monolith packing for the biomass growth and for assuring high mass transfer between gas and liquid phases since they seem to develop Taylor flow inside the reactor (Vedova 2008).

Two-phase partitioning bioreactors have been developed to treat hydrophobic compounds (Vedova 2008). Organic solvent can increase the retentivity of hydrophobic pollutants (Collins and Daugulis 1997). The stirring system allows a good mass transfer rate between gas, the aqueous, and the organic phases. The clogging or drying is avoided. The organic phase can also be solid, mainly polymer beads which have high affinity to oxygen (Daugulis et al. 2007).

Despite the wide variety of different bioreactors for waste gas treatment, conventional biofilters and biotrickling filters remain the most used equipments. The choice of the most suitable bioreactor depends on the characteristics and the composition of the waste gas and on the economical aspects as well. Biotrickling filters seem to be the best bioreactor, since the process can be easier controlled with better performances. However, the knowledge of the characteristics of every bioreactor is really very important to individuate the critical aspects of the process and to reduce the risk of malfunctioning.

References

Bielefeldt AR, Stensel HD (1999) Treating VOC-contaminated gases in activated sludge: mechanistic model to evaluate design and performance. Environ Sci Technol 33:3234–3240

Burgess JE, Parsons SA, Stuetz RM (2001) Developments in odour control and waste gas treatment biotechnology: a review. Biotechnol Adv 19(1):35–63

Collins LD, Daugulis AJ (1997) Biodegradation of phenol at high initial concentrations in two-phase partitioning batch and fed-batch bioreactors. Biotechnol Bioeng 55:152–162

Daugulis AJ, Littlejohns JV, Boudreau NG (2007) Solid–liquid two-phase partitioning for the treatment of gas-phase VOCs. In: Proceedings of the 2nd international congress biotechniques for air pollution control, A Coruˇna, Spain, 3–5 Oct 2007

De Bo I, Van Langenhove H, Pruuost P, Van De Steene I (2000) Oxygen supply and adhesion of monocultures in a membrane bioreactor. Toegepast Biol Wet 65(3a):63–69

De Bo I, Heyman J, Vincke J, Verstraete W, Van Langenhove H (2003) Dimethyl sulfide removal from synthetic waste gas using a flat poly(dimethylsiloxane)-coated composite membrane bioreactor. Environ Sci Technol 37:4228–4334

De Bo I, Jacobs P, Van Langenhove H (2002) Removal of toluene and trichloroethylene from waste air in a membrane bioreactor. Environ Sci Pollution Res 3:28 (Special Issue)

Jin Y, Veiga MC, Kennes C (2007) Behaviour and optimization of a novel monolith bioreactor for waste gas treatment. In: Proceedings of the 2nd international congress biotechniques for air pollution control, A Coruˇna, Spain, 3–5 Oct 2007

Kan E, Deshusses MA (2003) Development of foamed emulsion bioreactor for air pollution control. Biotechnol Bioeng 84(2):240–244

Kan E, Deshusses MA (2006) Cometabolic degradation of TCE vapors in a foamed emulsion bioreactor. Environ Sci Technol 40:1022–1028

Kennes C, Veiga MC (2001) Bioreactors for waste gas treatment. Kluwer Academic, Dordrecht. ISBN: 0-7923-7190-9

Kraakman JJR, van Ras N, Llewellyn D, Starmans D, Rebeyre P (2007) Biological waste gas purification using membranes: opportunities and challenges. In: Proceedings of the 2nd international congress biotechniques for air pollution control, A Coruˇna, Spain, 3–5 Oct 2007

Mota M, Peixoto J (2008) Inocula selection for VOC removal in the non-clogging biological plate tower. In: Ferreira EC, Mota M (eds) Book of abstracts of the 10th international chemical and biological engineering conference—CHEMPOR 2008, University of Minho, Braga, Portugal, pp 933–934, 4–6 Sept 2008

Neal AB, Loehr RC (2000) Use of suspended-growth reactors to treat VOCs. Waste Manag 20:59–68

Norman CW (2002) The effect of periodic operation on biofilters for removal of methyl ethyl ketone from contaminated air. Dissertation, Louisiana State University. http://etd.lsu.edu/docs/available/etd-0418102141608/.../Norman_thesis.pdf

Peixoto JM, Ribeiro Pinto JC (2012). Biofilm growth and hydrodynamic behaviour in the biological plate tower (BPT) with and without hanging biomass (BPT-HB). Water Sci Technol 66(8):1678–1683

Phipps DW (1998) Biodegradation of volatile organic contaminants from air using biologically activated foam, US Patent No. 5,714,379, February, 3

Reij MW, Keurentjes JTF, Hartmans S (1998) Membrane bioreactors for waste gas treatment. J Biotechnol 59:155–167

Sercu B, Peixoto J, Demeestere K, van Helst T, van Langenhove H (2006) Odors treatment: biological technologies. In: Nicolay X (ed) Odors in the food industry. Springer, Berlin, pp 125–158

Shareefdeen Z, Herner B, Singh A (2005) Biotechnology for air pollution control—an overview. In: Shareefdeen Z, Singh A (eds) Biotechnology for odor and air pollution control. Springer, Berlin

Song J, Shin S (2007) Biodegradation of BTXs and substrate interaction in a biocative foam reactor. In: Proceedings of the 2nd international congress biotechniques for air pollution control, A Coru˜na, Spain, 3–5 Oct 2007

Vedova LD (2008) Biofiltration of industrial waste gases in trickle-bed bioreactors. Case study: trichloroethylene removal. Scuoladi dottorato in ingegneria industriale indirizzo ingegneria chimica, XX Ciclo

Chapter 5
Removal of Odours

Biofiltration is capable of biodegrading a wide variety of air contaminants (Tonga and Skladany 1994; Bohn 1975; Prokop and Bohn 1985; Ottengraf and van den Oever 1983; Deshusses et al. 1995; Ottengraf 1986, 1987; Ottengraf et al. 1986; Thorsvold 2011; Brauer 1986; Cloirec et al. 2001; Hort et al. 2009; Chung 2007; Anit and Artuz 2000; Kosteltz et al. 1996; Yudelson 1996; Govind and Melarkode 1998; Govind et al. 1998). The process was initially applied to odour abatement in composting works, waste water treatment plants and similar situations. It is known that in 1953 a soil biofilter system was used for the treatment of odourous air in Long Beach, CA, USA (Pomery 1982). In Europe, the first attempt with a soil bed was made in Geneva for deodorization at a composting facility (Ottengraf 1986). Around 1959 a soil bed system was used at municipal sewage treatment in Nuremberg, Germany (Leson and Winer 1991; Shimko et al. 1988). In early 1960s, Carlson and Leiser (1966) started systematic research on biofiltration in the USA and used biofilters to treat hydrogen sulphide emissions from sewage. After that biological gas cleaning made considerable progress, but is still in its developing stages for application to the control of volatile organic compounds and air toxics in industrial use.

During the last 3 decades research activities, especially on the soil bed systems, have intensified in USA with the installation of some full-scale operations. Several researchers have published excellent reviews on the historical development of biofiltration (Ottengraf 1986; Leson and Winer 1991; Shimko et al. 1988). Having proven its success in deodorization, current research and application of biofiltration has been focused on the removal of volatile organic compounds and air toxics from the chemical and other process industrial exhausts. The research activities are currently focused on:

- Understanding the practical behaviour of the biofiltration process
- Optimizing its operational parameters
- Modelling the system on the basis of reaction kinetics for single- and the multiple-contaminant gas streams

Hydrogen sulphide is one of the most frequently produced odourous compounds in industrial processes like petroleum refining, rendering, wastewater treatment,

P. Bajpai, *Biological Odour Treatment*, SpringerBriefs in Environmental Science, DOI 10.1007/978-3-319-07539-6_5, © The Author(s) 2014

food processing and pulp and paper manufacturing. Therefore, its biofiltration has been studied extensively (Yang and Allen 1994a, b; Degorce-Dumas et al. 1997; Wada et al. 1986; Cook et al. 1999; Smet et al. 1996; Furusawa et al. 1984; Van Langenhove et al. 1986; Lee and Shoda 1989; Sercu 2006). The bacteria which are responsible for hydrogen sulphide degradation in biofilters generally belong to the genera *Thiobacillus* and *Acidithiobacillus* and are either neutrophilic or acidophilic. Hydrogen sulphide is oxidized to sulphuric acid under optimal conditions, but during stress conditions of high loads and oxygen limitation accumulation of elemental sulphur has been observed. Because hydrogen sulphide is very biodegradable, most investigations report very efficient hydrogen sulphide removal in a wide range of concentration. For example Yang and Allen (1994a, 1994b) reported higher than 99.9 % removal efficiencies for hydrogen sulphide inlet concentrations ranging from 5 to 2650 ppmv. However, because sulphuric acid is produced, acidification of the filter material will unavoidably occur during the biofiltration process. Its rate depends on the buffer capacity of the filter bed and the amount of hydrogen sulphide removed.

Degorce-Dumas et al. (1997) reported that buffering the packing to a near-neutral pH almost doubled the length of the period during which more than 95 % hydrogen sulphide removal efficiency was obtained. When the pH dropped below 6.6, the hydrogen sulphide removal efficiency started to decrease and the number of nonacidifying thiobacilli also decreased. The population of acidifying thiobacilli became dominant. For that reason, a correlation between the number of nonacidifying thiobacilli and the hydrogen sulphide removal efficiency was suggested. Other authors observed a smaller effect of acidic pH on the removal efficiency of hydrogen sulphide.

Yang and Allen (1994a, 1994b) observed almost equal hydrogen sulphide removal efficiencies at pH values between 3.2 and 8.8. Only at pH–1.6, the removal efficiency decreased to 15 %. The high hydrogen sulphide removal efficiency at pH–3.2 was ascribed to the abundance of acidophilic sulphur-oxidizing bacteria. Also other studies by Wada et al. (1986); Cook et al. (1999); Yang et al. (1994a) did not report decreased hydrogen sulphide removal efficiencies at pH values as low as 3 or even 1.2. Cook et al. (1999) reported that during biofiltration, the pH will first decrease at the inlet side of the biofilter, where most of the hydrogen sulphide is oxidized and the low pH front will therefore move to the deeper parts of the biofilter. In general, it should be sufficient to maintain a pH value higher than 3 for the efficient removal of hydrogen sulphide. However, it could be useful to maintain neutral pH values to avoid inhibition of the removal of other compounds present in the waste gas, corrosion and increased filter medium degradation. Yang and Allen (1994b) suggested that in order to increase the pH of the biofilter material, washing can be applied although only small pH increases are usually obtained. Smet et al. (1996) reported that regeneration of an acidified biofilter (pH–4.7) was not possible by trickling tap water or buffer solution over the bioreactor, as most of the sulphate was leached as the corresponding sulphate salts and not as sulphuric acid. In addition, leaching caused wash-out of essential microbial elements. Alternatively, the use of more concentrated buffer solutions in combination with a complete mineral medium or mixing with limestone powder was suggested.

Furusawa et al. (1984) used a packed bed of fibrous peat for the removal of hydrogen sulphide from air. Hydrogen sulphide was almost completely removed irrespective of its inlet concentration when the loading was less than 0.44 g sulphur/day/kg of dry peat. The removal rate of hydrogen sulphide by the acclimatized peat was fairly constant under a constant inlet concentration but the reaction rate constant was proportional to the influent concentration of hydrogen sulphide. Van Langenhove et al. (1986) reported the elimination of hydrogen sulphide from odourous air using a wood bark filter to improve the low permeability of soil beds. Lee and Shoda (1989) reported the biological removal of methyl mercaptan using an activated carbon fabric as a carrier of microorganisms for the biofilters. The activated carbon fabric seeded with digested night soil was found to be best packing material amongst the five materials evaluated. The critical load of methyl mercaptan, in which the gas can be completely removed, was determined as 0.48 g sulphur/kg of activated carbon fabric/day. About 80% of methyl mercaptan removed in the biofilter was converted into the sulphate ion. Effluent gas concentrations of methyl mercaptan and dimethyl disulphide were not detected below 50 ppm inlet concentration at a space velocity of 50/h. Fibrous materials which are flexible, light and less microbially degradable may become significant as carriers of microorganisms.

Hirai et al. (1990) studied the kinetics of removal of three kinds of odourous sulphur compounds—hydrogen sulphide, methanethiol and dimethyl sulphide—in acclimatized peat by supplying single or mixed odourous gases. Hydrogen sulphide and methanethiol were found to be degraded in peat irrespective of the acclimatizing gas, and their maximum removal rates were unaffected by the presence of dimethyl sulphide. Whereas, dimethyl sulphide was degraded only in dimethyl sulphide-acclimatized peat. It has been reported that the peat has the advantages over soil or compost of broadness of the maximum permeability of the moisture content and a lower pressure drop due to its fibrous structure. The same research group reported earlier about the characteristics of the peat as a packing material in deodorization device with the following results: zero-order kinetics in complete hydrogen sulphide removal by peat biofilters (Furusawa et al. 1984), characteristics of isolated hydrogen sulphide-oxidizing bacteria inhabiting a peat biofilter (Wada et al. 1986) and biological removal of organosulphur compounds by peat biofilters (Hirai et al. 1988). Gradual increase of load was found to be better for obtaining a high removal rate than the high load at the start of the experiment. Acclimation periods for hydrogen sulphide, methanethiol and dimethyl sulphide were 19, 17 and 24 days, respectively. The pH of the peat gradually decreased due to accumulation of sulphate ions during this period.

The maximum removal rate of hydrogen sulphide in its acclimatized peat was one order larger than those in methanethiol and dimethyl sulphide-acclimatized peat. The removal of dimethyl sulphide was affected by the mixed gasses. Although the removal of dimethyl sulphide decreased when present with methanethiol, the existence of hydrogen sulphide will weaken the effect of methanethiol on dimethyl sulphide removal to a certain extent. Thus, it would be better to maintain the space velocity value lower in order to guarantee dimethyl sulphide removal (Hirai et al. 1990). Two stage columns in series are recommended at a high space velocity. In the first column, most of the hydrogen sulphide and methanethiol can be removed, while the second column can be used exclusively for dimethyl sulphide removal. This method is also found to be suitable for the maintenance of operations including the washing of accumulated ions and the exchange of packing material.

While biooxidizing hydrogen sulphide and organic sulphur compounds in a filter, accumulation of sulphate can easily reach a level that can significantly reduce the biological activity of the biofilter. Therefore, sulphate should be periodically washed off before it reaches the toxic level. The removal of dimethyl sulphide decreases considerably if methanethiol is also present in the exhaust gas (Hirai et al. 1990). However, the existence of hydrogen sulphide weakens the effect of methanethiol on dimethyl sulphide removal rate to a certain extent. In this case, it would be desirable to maintain a low space velocity to ensure dimethyl sulphide removal. At high space velocity, two stage columns in series are recommended. So that, in the first column, most of the hydrogen sulphide and methanethiol can be removed, while the second column will be exclusively for dimethyl sulphide removal. This method may also be suitable for the maintenance of operations, including the washing of accumulated ions and the replacement of packing material. Multistage operation of biofilters may also be necessary when the waste gasses contain components which require different conditions for their microbial degradation. This way, optimal growth conditions for the different microbial population can be provided in separate stages. Also, more stages may be necessary when the waste gasses include one component in a concentration so high that the capacity of one stage is not enough for a sufficient degradation. Depending on the nature of the organic compounds present in the waste, the filter sometimes needs inoculation with suitable microorganisms to start biological activity.

An eight-membered bacterial consortium was used by Shareefdeen et al. (1993). This consortium was obtained from methanol-exposed soil, and a peat–perlite column for the biofiltration of methanol vapours. The biofilter was found to be effective in removing methanol at rates up to 112.8 g/h/m^3 packing. They also derived a mathematical model and validated it. Both experimental data and model predictions suggested that the methanol biofiltration process was limited by oxygen diffusion and methanol degradation kinetics. Bench-scale experiments and a numerical model were used by Hodge and Devinny (1994) to test the effectiveness of biofiltration in treating air contaminated with ethanol vapours. Out of the three different packing materials used—granular activated carbon, compost and a mixture of compost and diatomaceous earth—the granular activated carbon supported the highest elimination rates, ranging from 53 to 219 g/m^2/h for a range of loading rates. Partitioning coefficients for the contaminant on the biofilter packing material had a strong effect on the efficiency of the biofilters. Several studies on removal of volatile solvents like ketone mixtures, toluene, ethyl acetate by biofiltration have also been reported (Kirchner et al. 1987; Campbell and Connor 1997; Bibeau et al. 1997; Deshusses et al. 1997).

The performance of biofiltration to remove odours from an animal rendering plant's gaseous emissions was investigated by Luo and Oostrom (1997) using pilot-scale biofilters containing different media—sand, sawdust, bark and bark/soil mixture. Biofilter odour removal efficiencies of 75–99% were obtained at various air loading rates (0.074–0.057 m^3/m^3 medium/min) and medium moisture contents. Bio-Reaction Industries Inc., Tualatin, OR, USA has reported the development of a modular vapour-phase biofilter which is capable of treating extremely high concentrations of volatile organic compounds in low air volumes (Stewart and Thom 1997). These systems are more suitable for point source industrial process air streams, storage tanks and other vent emissions.

Biofiltration of NO_x is reported to be increased by the addition of an exogenous carbon and energy source (Apel et al. (1995). pH control is found to be an important operating parameter due to the acidic nature of the gas. The addition of calcite to the biofilter bed provided an effective internal buffer and the optimum temperature was found to be 50–60 °C. The biofilter using activated carbon or anthracite as the packing material was reported to be the most acceptable process for the removal of malodourous compounds containing nitrogen or sulphur because it produced no oxidized organics noticed with ozonation, and it had an equally high removal efficiency of both sulphur- and nitrogen-containing odourous compounds (Hwang et al. 1995).

Biofiltration has been successfully applied to remove α-pinene, a very hydrophobic volatile organic compound discharged in pulp and paper and wood products emissions, from a contaminated air stream (Mohseni and Grant 1997). Two identical bench scale biofilters were used for more than 4 months. The biofilter medium consisted of a mixture of wood chips and spent mushroom compost which was amended with higher perlite, for the first biofilter and with granulated activated carbon, for the second biofilter. The experiment was conducted at loading rates between 5 and 40 g α-pinene/m^3 bed medium/h. Under steady-state operating conditions, both biofilters, amended with perlite and granular activated carbon, performed similarly and provided removal rates of up to 30–35 g α-pinene/m^3 bed medium/h with gas retention times as low as 30 s. The adsorption characteristics of granular activated carbon were significant only during the start-up period, in which the granular activated carbon biofilter had a significantly better performance than perlite biofilter. When the biofilters were subjected to a sudden increase in the loading rate, the performance of the biofilters decreased significantly. The reacclimation period, however, was not long and biofilters reached more than 99% removal within less than 48 h of the spike load.

Deshusses (1997) studied the transient behaviour of a laboratory-scale compost-based biofilters. This included start-up, carbon balances and interactions between pollutants in the aerobic biodegradation of volatile organic compounds mixtures from effluent air streams. The study of transient behaviour offers a genuine basis for the development of a conceptual explanation of the complex phenomena that occur in biofilters during pollutant elimination, thereby providing an opportunity for further progress in establishing basic understanding of such reactors (Shareefdeen and Baltzis 1994; Tang et al. 1995; Deshusses et al. 1995). During long-term operation of a biofilter, the mandatory absence of net cell growth forces the cells into maintenance metabolism, which is of relatively low rate compared to substrate consumption during the active growth of the acclimation phase. Postacclimation nutrient addition increases activity primarily by allowing a return to the high substrate consumption rate of active growth, and only secondarily helps raise bed activity because of the ultimately higher amount of biomass in the bed (Cherry and Thompson 1997). The biomass content of a biofilter during the acclimation phase can be estimated using two approximate methods. The first follows the cumulative amount of substrate converted and uses the yield of cells from substrate during active growth to estimate the total biomass created. The second method follows a rate constant for conversion of substrate in the bed. This number is proportional to the amount of biomass as long as the conditions in the bed example temperature, pH and substrate concentration are relatively constant (Cherry and Thompson 1997).

Generally, the empirical knowledge dictates the design and scale-up of biofiltration plants, even though substantial performance improvement could be expected from a more comprehensive knowledge of the individual processes involved in pollutant elimination. For improved design and performance, an appropriate model for the whole process is required. Deshusses (1997) and Deshusses et al. (1995) have developed a novel diffusion reaction model for the determination of both the steady-state and transient-state behaviour of biofilters for waste air treatment, and experimentally evaluated/verified the same. Although this model deals with the aerobic biodegradation of methyl ethyl ketone and methyl isobutyl ketone vapours from air, similar mathematical treatment can be given to other biofilters degrading hydrogen sulphide, organosulphur compounds, and other volatile organics. Most of the mathematical models have been developed mainly to correlate a particular set of experimental data, to explain the influence of selected parameters on the efficiency of the process, and sometimes to seek a better basic understanding of the phenomena occurring in a biofilter (Shareefdeen et al. 1993; Hodge and Devinny 1994; Deshusses et al. 1995). Choi et al. (1996) has presented a more promising quantitative structure–activity relationship for biofiltration.

Qiao et al. (2008) studied the removal characteristics of hydrogen sulphide experimentally in the biofilters. They used fibrous peat and resin as the packed materials. The biofilter with 100 % of the peat showed higher removal capacity in comparison to resin biofilter, but the gas flow resistance was lower in the latter. The mixture of the peat and resin as the packed material of the biofilter was proved to be an advisable method to keep the high removal capacity and reduce the gas flow resistance for a long-term operation. The flow resistance can decrease by 50 % when 50 % of the resin mixed with the peat, but the removal capacity was still considerably higher.

Wani et al. (2001) studied biofiltration using compost and hog and a mixture of the two to remove reduced sulphur (RS) gases emitted from pulp mills. The hog fuel exhibited more resistance to microbially induced bed degradation in comparison to compost or mixtures of both, and was found to be effective at RS gas removal as compost, with the advantage of costing less.

Biological filtration oxygenated reactor (Biofor) is a new generation of modern apparatus, an aerobic biological reactor from Degremont, with fixed biomass on a support material (Brenna 2000). The main advantages of biofiltration are a high concentration of biomass that brings the reactor to operation without the problems of bulking, with the elimination of pollutants difficult to degrade biologically. Biofor gives these results as a result of an ideal support material, efficient aeration system, a process of ascending equal currents of air and water and optimized washing processes. The support material, Biolite, presents optimal qualities of density, hardness, friction and porosity. As well as working without odours and noise, Biofor is adapted for plants to limit environmental impact.

Few researchers reported successful removal of organic sulphur compounds in a full-scale biofilter treating emissions from mushroom composting, after inoculation with a specialized strain (Sercu et al. 2006). Fifty days after inoculation, the total sulphur removal efficiency in the inoculated biofilter section had increased to 99% compared with 68% in the noninoculated section. But even when inoculation is used, in a mixture of RS compounds, hydrogen sulphide is preferentially degraded over dimethyl sulphide or other organic sulphur compounds (Cho et al. 1992; Wani et al. 1999; Zhang et al. 1991). This occurs because hydrogen sulphide oxidation yields most energy for the microorganisms (Smet et al. 1996). Therefore, the bioreactor has to be designed large enough to allow hydrogen sulphide degradation at the inlet side of the biofilter and degradation of the remaining volatile organic sulphur compounds deeper in the biofilter bed. Finally, when a biofilter is designed properly to remove volatile organic sulphur compounds, there is still a change of long-term decrease in removal efficiency because of acidification. Similarly as for hydrogen sulphide, sulphuric acid is formed after complete oxidation of volatile organic sulphur compounds. Microorganisms degrading the volatile organic sulphur compounds, however, are much more sensitive to low pH values than hydrogen sulphide-oxidizing bacteria. Smet et al. (1996), for instance, observed a decreased dimethyl sulphide elimination capacity when the compost pH decreased below 5. To prevent problems due to acidification, the bioreactor has to be designed large enough, and for high influents loadings, pH control should be included. Alternatively, two-stage systems have been proposed, first removing hydrogen sulphide and subsequently volatile organic sulphur compounds (Kasakura and Tatsukawa 1995; Park et al. 1993; Ruokojarvi et al. 2001; Sercu et al. 2005b).

Biotrickling filters have been also studied for removal of hydrogen sulphide. Their main advantages are optimal control of pH, nutrients and accumulation products, but the treatment costs are higher. At an empty bed residence time between 30 and 120 s, high hydrogen sulphide removal efficiencies, more than 95% easily can be obtained for hydrogen sulphide concentrations between 200 and 2000 ppmv (Ruokojärvi et al. 2001; Sercu et al. 2013b). At lower influent concentrations, lower empty bed residence time can be used at high removal efficiencies. Gabriel and Deshusses (2003) described the retrofitting of existing chemical scrubbers for hydrogen sulphide removal to biotrickling filters, maintaining an empty bed residence time between 1.6 and 2.2 s. Removal efficiencies of more than 98% were commonly reached for 30 ppmv inlet concentrations, with decreases to 90% at 60 ppmv peak concentrations. The removal of volatile organic sulphur compounds in the same reactor was lower however, for example, 35±5% for carbon disulphide. The authors attributed the residual odour after the biotrickling filter mainly to the persistence of these compounds. Also Wu et al. (2001) obtained more than 95% dimethyl sulphide removal efficiency at empty bed residence time − 5 s, at less than 6 ppmv influent concentrations in a pilot-scale biotrickling filter. At 20 ppmv influent concentration, the removal efficiency reduced to about 89%.

Van Langenhove et al. (1992) made a comparison of a full-scale biotrickling filter and a biofilter for treating rendering emissions. Both techniques were found to remove alkanals very efficiently, but organic sulphur compounds were much less efficiently removed. This was ascribed to an insufficient development of microorganisms capable of degrading these compounds. Goodwin et al. (2000) also observed problems removing RS compounds with a biofilter at a biosolids composting facility. Increasing the empty bed residence time from 20 to 32 s improved the removal efficiency somewhat. In contrast, volatile organic compounds like methane, formaldehyde, isopentanal, N,N-dimethylmethenamine, and dimethylamine were removed for more than 95% in all cases at average inlet concentrations of 15 ppmv.

Goncalves and Govind (2010) treated hydrogen sulphide-polluted airstreams in two biotrickling filter columns packed with polyurethane foam cubes, one column with cubes coated with a solution of 25 mg/L of polyethyleneimine (coated reactor) and the other containing just plain polyurethane cubes (uncoated reactor) at empty bed residence times ranging from 6 to 60 s and inlet hydrogen sulphide concentrations ranging from 30 to 235 ppmv (overall loads of up to 44 g hydrogen sulphide/ m^3 bed/h), with overall removal efficiencies in the range of 90−100% over 125 days. The acclimatization characteristics of the coated reactor outperformed those of the uncoated one, and both the observed elimination capacity of 77 g hydrogen sulphide/m^3 bed/h and retention of volatile solids of 42 mg volatile solids/cube were maximum in the coated reactor. Insights into the controlling removal mechanisms were also provided by means of dimensionless analysis of the experimental data. Denaturing gradient gel electrophoresis showed that the dominant surviving species in both units belonged to the genus *Acidithiobacillus*.

Ruokojarvi et al. (2001) developed a two-stage biotrickling filter for sequential removal of hydrogen sulphide, methyl mercaptan and dimethyl sulphide. Two bioreactors connected in series were inoculated with enriched activated sludge, the first operating at low pH for removal of hydrogen sulphide and the second at neutral pH for removal of dimethyl sulphide. Methyl mercaptan was removed in both reactors. Hydrogen sulphide, dimethyl sulphide and methyl mercaptan elimination capacities (as sulphur) as high as 47.9, 36.6 and 2.8 $g/m^3/h$, respectively, were obtained for the entire two-stage biotrickling filter at more than 99% removal efficiencies and the reactor showed a good long-term stability.

Two *Hyphomicrobium* VS inoculation protocols were compared for start-up of a biotrickling filter removing dimethyl sulphide (Sercu et al. 2005a). A dynamic model was developed that described the removal of dimethyl sulphide in the presence of methyl alcohol in inorganic biofilters under both steady and transient conditions (Zhang et al. 2007a, 2007). Biological treatment of dimethyl sulphide was investigated in a bench-scale biofilter, packed with compost along with wood chips, and enriched with dimethyl sulphide-degrading microorganism *Bacillus sphaericus*

(Giri et al. 2010). Dimethyl sulphide was removed in a thermophilic biotrickling filter operated at 52 °C, using an enriched sludge inoculum (Luvsanjamba et al. 2008). The membrane bioreactor contained a polydimethylsiloxane/Zirfon composite membrane and inoculated with *Hyphomicrobium* VS, a methylotrophic microorganism, was used to remove dimethyl sulphide from waste air (Bo et al. 2002). The biofilter process and bacterial community composition are key elements for biodegradation of dimethyl sulphide. Hydrogen sulphide, methanethiol, dimethyl sulphide and dimethyl disulphide were degradated by *Hyphomicrobium* DW44 isolated from peat biofilter (Cho et al. 1991). Dimethyl sulphide was conversed by methylophagasulfidovoran in a microbial mat (Zwart and Kuenen 1997). A PCR-DGGE approach and a dendrogram had been used to illustrate the diversity of the bacterial community in a biofilter at different operating conditions. The diversity of the bacterial community in the biofilter is dynamic and varies with inlet dimethyl sulphide loads, the addition of glucose and fluctuating temperature (Chung et al. 2010). Wei et al. (2013) conducted experimental investigations to remove the odour-containing dimethyl sulphide in biofilter filled with the ceramsite as a medium. The biotrickling filter packed with ceramsite was set up to study the removal of dimethyl sulphide. The removal efficiency in the biotrickling filter was up to 99% based on experimental results. The optimal spray density, empty bed residence time and pH were 100 mL/min, 38 s and 6.0, separately. The microbial community composition taken from packing material samples in the biotrickling filter for removal of dimethyl sulphide developed, which were assessed by polymerase chain reaction-denaturing gradient gel electrophoresis of eubacterial 16S rDNA followed by clone library analysis, revealed four distinct bands. Phylogenetic analysis showed that the sequences of these bands were closest to sequences of species of the *Bacillus* genus, Rhodobacteraceae bacterium, proteobacterium and delta proteobacterium.

Domtar's kraft mill, Cornwall, Ontario, Canada carried out research for reducing the odours from the plant (Lau et al. 2006). Three types of biofiltration technology were examined: biofilters, bioscrubbers and biotrickling reactors. This last option seemed the most favourable for treating the gas leaving the brownstock reactor. With a biotrickling reactor conditions such as temperature, pH and growth of the biomass can be controlled. Four types of packing material were tried. The packing material should have a high void fraction, high specific surface area, be made from an acid-resistant material, have a low bulk density and the microorganisms should stick to the packing. Lantec's HD Q-PAC gave the optimum results.

The biofiltration of dimethyl sulphide in simulated waste gas has been reported in the literature showing variation in the performance of the system (Giri et al. 2010; Chan 2006; Delhomenie and Heitz 2005; Shareefdeen et al. 2005). Reduced sulphurous compounds biofiltration generates acid, which reduces the pH of the packing medium thereby affecting the biodegradation (Shareefdeen et al. 2005; Christen et al. 2002; Maestre et al. 2007). In traditional biofilters without water recirculation, by-products of degradation of reduced sulphur compounds drop the pH of the biofilters. Some of the researchers have reported dimethyl sulphide degradation using methanol for cometabolism with improved biodegradation of dimethyl sulphide (Zhang et al. 2006, 2007a, 2007b, 2008; Darracq et al. 2010). Giri and

Pandey (2013) treated the ambient air and live vent gas from a pulp and paper industry containing dimethyl sulphide along with other traces of reduced sulphurous compounds in a biofilter packed with wood chips and compost, and seeded with the microorganism *B. sphaericus*. It was observed that a bench-scale biofilter packed with compost and wood chips seeded with potential dimethyl sulphide degrading culture (*B. sphaericus*) could efficiently remove dimethyl sulphide from ambient air with removal efficiency of 71 ± 11 at an effective bed contact time of 360 ± 20 s with loading rate in the range of 4–28 g dimethyl sulphide/m^3/h. Further, the same biofilter operated for the treatment of vent gas generated from a pulp and paper industry indicated dimethyl sulphide removal of $61 \pm 18\%$ at optimal effective bed contact time of 360 ± 25 s with a loading rate in the range of 3–128 g dimethyl sulphide/m^3/h.

References

Anit S, Artuz R (2000) Biofiltration of air. Rensselaer Institute. www.rensselaer.edu/dept/chem-eng/Biotech-Environ/MIS/biofilt/biofiltration.htm

Apel WA, Barnes JM, Barrett KB (1995) Biofiltration of nitrogen oxides from fuel combustion gas streams. In: Proceedings of 88th Annual Meeting of the Air and Waste Management Association, San Antonio, TX, 18–23 June 1995

Bibeau L, Kiared K, Leroux A, Brzezinski R, Viel G, Heitz M (1997) Biological purification of exhaust air containing toluene vapor in a filter-bed reactor. Can J Chem Eng 75:921–929

De Bo I, Langenhove HV, Heyman J (2002) Removal of dimethyl sulfide from waste air in a membrane bioreactor. Desalination 148:281–287

Bohn HL (1975) Soil and compost filters for malodorant gases. J Air Waste Manag Assoc 25:953–955

Brauer H (1986) Biological purification of waste gases. Int Chem Eng 26(3):387–395

Brenna V (2000) Biofor: biofiltration of paper mill effluents. Ind Carta 38(5):59–63

Carlson DA, Leiser CP (1966) Soil bed for control of sewage odors. J Water Pollut Control Fed 38:829–833

Campbell HJ, Connor MA (1997) Practical experience with an industrial biofilter treating solvent vapor loads of varying magnitude and composition. Pure Appl Chem 69(11):2411–2424

Chan AA (2006) Attempted biofiltration of reduced sulfur compounds from a pulp and paper mill in Northern Sweden. Environ Prog 25:152–160

Cherry RS, Thompson DN (1997) Shift from growth to nutrient-limited maintenance kinetics during biofilter acclimation. Biotechnol Bioeng 56(3):330–339

Cho KS, Hirai MI, Shoda M (1991) Degradation characteristics of hydrogen sulfide, methanethiol, dimethyl sulfide and dimethyl disulfide by *Hyphomicrobium* DW44 isolated from peat biofilter. J Fermen Bioeng 171:384–389

Cho KS, Hirai M, Shoda M (1992) Enhanced removability of odorous sulphur-containing gases by mixed cultures of purified bacteria from peat biofilters. J Ferment Bioeng 73:219

Choi DS, Webster TS, Chankg AN, Devinny JS (1996) Quantitative structure–activity relationships for biofiltration of volatile organic compounds. In: Reynolds Jr FE (ed) Proceedings of 1996 conference on biofiltration. The Reynolds Group, Tustin, pp 231–238

Christen P, Domenech F, Michelena G, Auria R, Revah S (2002) Biofiltration of volatile ethanol using sugar cane bagasse inoculated with *Candida utilis*. J Hazard Mater 89:253–265

Chung YC (2007) Evaluation of gas removal and bacterial community diversity in a biofilter developed to treat composting exhaust gases. J Hazard Mater 144:377–385

Chung YC, Cheng CY, Chen TY, Hsu JS, Kui CC (2010) Structure of the bacterial community in a biofilter during dimethyl sulfide (DMS) removal processes. Bioresour Technol 101:7176–7179

Cloirec PL, Humeau P, Ramirez-Lopez EM (2001) Biotreatments of odours and performances of a biofilter and a bioscrubber. Water Sci Tech 44(9):219–226

Cook LL, Gostomski PA, Apel WA (1999) Biofiltration of asphalt emissions: full-scale operation treating off-gases from polymer-modified asphalt production. Environ Prog 18:178

Darracq G, Couvert A, Couriol C, Amrane A, Cloriec PL (2010) Kinetics of toluene and sulfur compounds removal by means of an integrated process involving the coupling of absorption and biodegradation. J Chem Technol Biotechnol 85:1156–1161

Degorce-Dumas JR, Kowal S, Le Cloirec P (1997) Microbiological oxidation of hydrogen sulphide in a biofilter. Can J Microbiol 43:264

Deshusses MA (1997) Biological waste air treatment in biofilters. Curr Opin Biotechnol 8:335–339

Delhomenie MC, Heitz M (2005) Biofiltration of air: a review. Crit Rev Biotechnol 25:53–72

Deshusses MA, Hamer G, Dunn IJ (1995) Behavior of biofilters for waste air biotreatment. I: dynamic model development. Envir Sci Technol 29(4):1048–1058

Deshusses MA, Johnson CT, Hohenstein GA, Leson G (1997) Treating high loads of ethyl acetate and toluene in a biofilter. In: Proceedings of the Air & Waste Management Association 90th annual meeting and exhibition, 8–13 June 1997, Toronto, Canada, p 13

Furusawa N, Togashi I, Hirai M, Shoda M, Kubota H (1984) Removal of hydrogen sulfide by a biofilter with fibrous peat. J Ferment Technol 62(6):589–594

Gabriel D, Deshusses MA (2003) Retrofitting existing chemical scrubbers to biotrickling filters for H$_2$S emission control. Proc Natl Acad Sci U S A 100:6308

Giri BS, Pandey RA (2013) Biological treatment of gaseous emissions containing dimethyl sulphide generated from pulp and paper industry. Bioresour Technol 142:420–427

Giri BS, Mudliar SN, Deshmukh SC, Banerjee S, Pandey RA (2010) Treatment of waste gas containing low concentration of dimethyl sulphide (DMS) in a bench-scale biofilter. Biores Tech 101:2185–2190

Goncalves JJ, Govind R (2010) Enhanced biofiltration using cell attachment promotors. Environ Sci Technol 43(4):1049–1054

Goodwin JP, Amenta SA, Delo RC, Del Vecchio M, Pinnette JR, Pytlar TS (2000) Odor control advances at cocomposting facility. Biocycle 41:68

Govind R, Melarkode R (1998) Pilot-scale test of the biotreatment of odors from Zimpro^{TM} sludge conditioning process, Report submitted to sanitation District No. 1, Fort Wright, KY, by PRD TECH, Inc., Florence, KY

Govind R, Fang J, Melarkode R (1998) Biotrickling filter pilot study for ethanol emissions control, a report prepared for the food manufacturing coalition for innovation and technology transfer, by PRD TECH, Inc., Florence, KY

Hirai M, Ohtake M, Soda M (1990) Removal of kinetics of hydrogen sulfide, methanethiol and dimethyl sulfide by peat biofilters. J Ferment Bioeng 70:334–339

Hirai M, Terasawa M, Inamura I, Fujie K, Shoda M, Kubota H (1988) Biological removal of organosulfur compounds using peat biofilter. J Odor Res Eng 19:305–312

Hodge DS, Devinny JS (1994) Biofilter treatment of ethanol vapors. Environ Progr 13(3):167–173

Hort C, Gracy S, Platel V, Moynault L (2009) Evaluation of sewage sludge and yard waste compost as a biofilter media for the removal of ammonia and volatile organic sulfur compounds (VOSCs). Chem Eng J 152:44–53

Hwang Y, Matsuo T, Hanaki K, Suzuki N (1995) Identification and quantification of sulfur and nitrogen containing odorous compounds in waste water. Water Res 29(2):711–718

Kasakura T, Tatsukawa K (1995) On the scent of a good idea for odour removal. Water Qual Int 2:24

Kirchner K, Hauk G, Rehm HJ (1987) Exhaust gas purification using immobilized monocultures (biocatalyst). Appl Microbiol Biotechnol 26:579–587

Kosteltz AM, Finkelstein A, Sears G (1996) What are the 'real opportunities' in biological gas cleaning for North America? In: Proceedings of the 89th Annual Meeting and Exhibition of Air and Waste Management Association, A and WMA, Pittsburgh, PA; 96-RA87B.02

Lau S, Groody K, Chan A (2006) Control of reduced sulphur and VOC emissions via biofiltration. Pulp Pap Can 107(12):57–63

Leson G, Winer AM (1991) Biofiltration: an innovative air pollution control technology for VOC emissions. J Air Manag Assoc 41:1045–1054

Lee S-K, Shoda M (1989) Biological deodorization using activated carbon fabric as a carrier of microorganisms. J Ferment Bioeng 68(6):437–442

Luo J, van Oostrom A (1997) Biofilters for controlling animal rendering odor—a pilot-scale study. Pure Appl Chem 69(11):2403–2410

Luvsanjamba M, Sercu B, Peteghem JV, Langenhove HV (2008) Long-term operation of a thermophilic biotrickling filter for removal of dimethyl sulfide. Chem Eng J 142:248–255

Maestre JP, Gamisans X, Gabriel D, Lafuente J (2007) Fungal biofilters for toluene biofiltration: evaluation of the performance with four packing materials under different operating conditions. Chemosphere 67:684–692

Mohseni M, Grant AD (1997) Biofiltration of α-pinene and its application to the treatment of pulp and paper air emissions. Tappi Environ Conf Exhib 2:587–592

Ottengraf SPP (1986) Exhaust gas purification. In: Schonborn W (vol ed), Rehm H-J, Reed G (series eds) Biotechnology (microbial degradations), vol 8, VCH, Weinheim, Chap. 12, 425–452

Ottengraf SPP (1987) Biological system for waste gas elimination. Trends Biotechnol 5:132–136

Ottengraf SPP, Van Denoever AHC (1983) Kinetics of organic compound removal from waste gases with a biological filter. Biotechnol Bioeng 25:3089–3102

Ottengraf SPP, Meesters JPP, van den Oever AHC, Rozema HR (1986) Biological elimination of volatile xenobiotic compounds in biofilters. Bioprocess Eng 1:61–69

Park SJ, Hirai M, Shoda M (1993) Treatment of exhaust gases from a night soil treatment plant by a combined deodorization system of activated carbon fabric reactor and peat biofilter inoculated with *Thiobacillus thioparus* DW44. J Ferment Bioeng 76:423

Pomeroy D (1982) Biological treatment of odorous air. J Water Pollut Control Fed 54:1541–1545

Prokop WH, Bohn HL (1985) Soil bed systems for control of rendering plant odors. J Air Waste Manag Assoc 35:1332–1338

Qiao S, Fu L, Chi Y, Yan N (2008) Removal characteristics of hydrogen sulfide in biofilters with fibrous peat and resin. In: Proceedings of the 2nd international conference on bioinformatics and biomedical engineering, 2008. ICBBE 2008, pp 39963–39999

Ruokojärvi A, Ruuskanen J, Martikainen PJ, Olkkonen M (2001) Oxidation of gas mixtures containing dimethyl sulfide, hydrogen sulfide, and methanethiol using a two-stage biotrickling filter. J Air Waste Manag Assoc 51:11

Sercu B, Núñez D, Aroca G, Boon N, Verstraete W (2005a) Inoculation and start-up of a biotrickling filter removing dimethyl sulfide. Chem Eng J 113:127–134

Sercu B, Núñez D, Van Langenhove H, Aroca G, Verstraete W (2005b) Operational and microbiological aspects of a two-stage biotrickling filter removing hydrogen sulfide and dimethyl sulfide. Biotechnol Bioeng 90:259

Sercu B, Peixoto J, Demeestere K, van Helst T, van Langenhove H (2006) Odors treatment: biological technologies. In: Nicolay X (ed) Odors in the food industry. Springer, Berlin, pp 125–158

Shareefdeen Z, Baltzis BC (1994) Biofiltration of toluene vapor under steady-state and transient conditions-theory and experimental results. Chem Eng Sci 49:4347–4360

Shareefdeen Z, Herner B, Singh A (2005) Biotechnology for air pollution control—an overview. In: Shareefdeen Z, Singh A (eds) Biotechnology for odor and air pollution control. Springer, Berlin

Shareefdeen Z, Baltzis BC, Oh YS, Bartha R (1993) Biofiltration of methanol vapor. Biotechnol Bioeng 41:512–524

Shimko IG, Spasov VA, Chinennaya SK, Zakirova RI, Tananina IN, Perchugor GY, Pavlova OI (1988) Biochemical methods of freeing gas–air mixture from sulfur containing compounds. Fiber Chem 19:373–378

Smet E, Van Langenhove H, Verstraete W (1996) Long-term stability of a biofilter treating dimethyl sulphide. Appl Microbiol Biotechnol 46:191

Stewart WC, Thom RC (1997) High VOC loading in biofilters industrial applications. In: Emerging solutions VOC air toxics control: proceedings of a specialty conference sponsored by the Air & Waste Management Association, Pittsburgh, PA, pp 38–65

Tang HM, Hwang SJ, Hwang SC (1995) Dynamics of toluene degradation in biofilters. Hazard Waste Hazard Mater 12:207–219

Thorsvold BR (2011) Biological odor control systems: a review of current and emerging technologies and their applicability. nfo.ncsafewater.org/shareddocuments/websitedocuments/annual-conference/AC2011Papers/WW_T.pm_03.45_Thorsvold.pdf. Accessed 20 March 2014

Tonga AP, Skladany GJ (1994) Field pilot-scale vapor-phase treatment of styrene using biofiltration. In: Flathman PE, Jerger DE, Exner JH (eds) Bioremediation: field experience. Lewis, Ann Arbor, pp 507–521

Van Langenhove H, Wuyts E, Schamp N (1986) Elimination of hydrogen sulfide from odorous air by a wood bark biofilter. Water Res 20:1471–1476

Van Langenhove H, Bendinger B, Oberthür R, Schamp W (1992) Organic sulfur compounds: persistant odourants in the biological treatment of complex waste gases. In: Dragt AJ, van Dam J (eds) Biotechniques for air pollution abatement and odour control policies. Elsevier, Amsterdam, pp 309–313

Wada A, Shoda M, Kubota H, Kobayashi T, Katayama FY, Kuraishi H (1986) Characteristics of H_2S oxidizing bacteria inhabiting a peat biofilter. J Ferment Technol 64:161–167

Wani AH, Lau AK, Branion RMR (1999) Biofiltration control of pulping odors—hydrogen sulfide: performance, macrokinetics and coexistence effects of organo-sulfur species. J Chem Technol Biotechnol 74:9

Wani AH, Branion RMR, Lau AK (2001) Biofiltration using compost and hog fuel as a means of removing reduced sulphur gases from air emissions. Pulp Paper Can 102(5):27–32

Wei Z, Liu X, Ma C, He J, Huang Q (2013) Gas dimethyl sulfide removal in biotrickling filtration. Open Access Sci Rep 2:702 doi:10.4172/scientificreports.702

Wu L, Loo YY, Koe LCC (2001) A pilot study of a biotricking filter for the treatment of odorous sewage air. Water Sci Technol 44:295

Yang Y, Allen ER (1994a) Biofiltration control of hydrogen sulfide 1. Design and operational parameters. J Air Waste Manag Assoc 44:863–868

Yang Y, Allen ER (1994b) Biofiltration control of hydrogen sulfide 2. Kinetics, biofilter performance, and maintenance. J Air Waste Manag Assoc 44:1315

Yudelson JM (1996) The future of the U.S. biofiltration industry. In Reynolds FE (ed) Proceedings of the 1996 conference on biofiltration (an air pollution control technology). The Reynolds Group, Tustin, p 1

Zhang L, Hirai M, Shoda M (1991) Removal characteristics of dimethyl sulfide, methanethiol and hydrogen sulfide by *Hyphomicrobium* sp. I55 isolated from peat biofilter. J Ferment Bioeng 72:392

Zhang Y, Liss SN, Allen DG (2007a) Enhancing and modeling the biofiltration of dimethyl sulfide under dynamic methanol addition. Chem Eng Sci 62:2474–2481

Zhang Y, Liss SN, Allen DG (2007b) Effects of methanol on pH and stability of inorganic biofilters treating dimethyl sulfide. Environ Sci Technol 41:3752–3757

Zhang YF, Liss SN, Allen DG (2006) The effects of methanol on the biofiltration of dimethyl sulfide in inorganic biofilters. Biotechnol Bioeng 95:734–743

Zhang YF, Allen DG, Liss SN (2008) Modeling the biofiltration of dimethyl sulfide in the presence of methanol in inorganic biofilters at steady state. Biotechnol Prog 24:845–851

Zwart JMM, Kuenen JG (1997) Aerobic conversion of dimethyl sulfide and hydrogen sulfide by *Methylophaga sulfidovorans*: implications for modeling DMS conversion in a microbial mat. FEMS Microbiol Eco 22:155–165

Chapter 6
Future Prospects

Biofiltration is an emerging technology for the control of odourous emissions. It is also energy efficient and has been used extensively in the USA and Europe for the control of odours from wastewater treatment facilities, rendering plants, composting facilities and other odour-producing operations (Adler 2001; Govind and Bishop 1996; Thorsvold 2011; Soccol et al. 2003; Shareefdeen et al. 2005; Gerrard et al. 2000). During the past few years, it has been used increasingly in the USA for treating high-volume, low-concentration air streams. Numerous research studies are being conducted to characterize its suitability for a wide variety of air emission control applications. This technology has proved to be a valuable alternative to the physiochemical treatment systems in odour abatement in both developed and developing countries.

Historically, biofiltration has been most commonly applied to remove odourous compounds such as hydrogen sulphide from air emissions at wastewater treatment plants (Brauer 1986). Since the 1980s, however, biofiltration has also been used to eliminate volatile organic compounds in gases emitted from a wide range of processes (van Groenestijn and Hesselink 1995; Leson and Winer 1991). This technology is attractive for many reasons including its ability to convert pollutants into inert products such as carbon dioxide and hydrogen peroxide at ambient temperatures. Another advantage of biofilters is that they do not generate secondary contaminant problems and thus are an environment friendly treatment method. Biofilters and biotrickling filters can be a more cost-effective option than the conventional air pollution control methods for high-volume, low-concentration gas streams containing readily biodegradable contaminants. It is efficient at ambient atmospheric conditions of temperature, pressure, pH, moisture and oxygen requirement (Van Lith et al. 1997). Finally, because these systems operate at ambient temperatures and do not require high-temperature media regeneration systems, they have lower energy requirements than competing technologies. The microbial flora survive a fairly long period during which the filter bed is not loaded (periods of a fortnight are easily spanned with hardly any loss of microbial activity). This is important in view of the dynamic behaviour of filter bed at discontinuous operation, and means a very short starting time after longer periods of not operating the filter bed (Ottengraph and Van Denoever 1983). Moreover, the presence of a large amount of packing material

P. Bajpai, *Biological Odour Treatment,* SpringerBriefs in Environmental Science, DOI 10.1007/978-3-319-07539-6_6, © The Author(s) 2014

with a buffering capacity diminishes the sensitivity of biofilters to different kinds of fluctuations.

Biological removal of odours is becoming more common as the experience and confidence in these technologies increase. The capital costs of biological treatment will continue to become more competitive with carbon and chemical scrubbers on a capital cost basis as the research continues and the technology advances and develops ways to improve the removal efficiencies. As the capital cost gap narrows, this will result in biological technologies being selected more often based on life-cycle cost over competing technologies. The continued optimization of biological systems for what they do best and combining with other technologies to address their limitations will also serve to promote the biological technologies. Thorsvold (2011) has summarized the key advantages and limitations of the different treatment technologies (Table 6.1).

Research continues on biological systems in the private and public sectors. New medias and concepts are being developed and investigated in order to produce higher loading rates to reduce the cost and increase the removal efficiencies. One interesting development introduced in the USA is the Bord Na Mona Monashell™ biotrickling filters (Thorsvold 2011). This system utilizes ordinary clam shells as the media for bacterial growth in a biotrickling filters configuration. This technology was patented in 1996 and has more than 600 installations worldwide, but it has only been used and tested in the USA. Studies have shown that these clam shells have high porosity, low differential pressure and an affinity for sulphur compounds; these properties make the shells a good choice as a biotrickling filters medium (Naples 2010). The naturally occurring calcium carbonate in the shells serves as a buffer to maintain a neutral pH in the biotrickling filters which allows the media bed to contain both autotrophic and heterotrophic bacteria throughout the biotrickling filters. The heterotrophic bacteria require a neutral pH to thrive. The autotrophic bacteria does the same in neutral pH as they do in an acidic environment. Studies have shown that the clam shell based biotrickling filters can remove high levels of hydrogen sulphide and organic reduced sulphur compounds, making this a potentially complete solution for a plant in a sensitive area and a desire to treat odours biologically without a carbon polishing stage. Results showed hydrogen sulphide removal efficiencies exceeded 99 %, even during spikes of almost 400 ppm hydrogen sulphide (Naples 2010), while also providing high removal efficiencies of organic reduced sulphur compounds. The problem with this media is that the calcium carbonate is consumed and the clamshell media must be replaced when it breaks down and collapses to the point where the media shows excessive head loss. In this way, the unit functions similarly to an organic media biofilter, though at much higher loading rates and with slower degradation of the media. Like the organic biofilters, the higher the hydrogen sulphide concentrations, the more frequently the media will need replacement.

Biofiltration will play a major role in the treatment of organic and inorganic emissions from a variety of industrial and waste water treatment processes. The applicability of the three types of biofilters—conventional biofilter, biotrickling filter and bioscrubber—depends to a large extent on the waste gas characteristics such as its solubility, biodegradability and the potential formation of acidic intermediates products. Compost biofilters are better suited for treatment of odours and low concentration (<25 ppmv) contaminants. Biotrickling filters have significant advantages over compost biofilters and are capable of handling significantly higher contaminant concentrations (20–5,000 ppmv) (Govind and Bishop 1996). The major issues in biotrickling filters are the design of the support media and handling of biomass growth. Support media design has a significant impact on biotrickling filter performance. The market for biofilters will increase in the next millennium, as new applications arise in the future.

Several ongoing trends in the development of biofiltration can be noted. Increasing biodegradation rates, particularly for less biodegradable organics, by introducing appropriate *microorganisms* and improving their environmental conditions is of high priority since it allows reductions in the required filter size and makes biofiltration an even more competitive air pollution control technology. At the same time, further improvements to the physical properties and longevity of the filter material are needed because they will result in reduced cost for energy and maintenance. Finally, full control of operating parameters allows further reduction in maintenance requirements and reduces the likelihood of system upsets.

Although such methods have long been known to be cost-effective, they have not found general acceptance in practice, even when the exhaust gas components to be removed are biodegradable. Long adaptation periods of the biomass in particular with large exhaust gas flow discontinuities or low space velocities i.e., low specific purification capacities, are the reasons often cited. Bed compaction problems, particularly with soil and compost biofilters, have also been noticed. This results in high pressure drop across the filter. However, with the help of granulated activated carbon and other synthetic packing materials, individually or in combination with soil-peat-compost materials, have solved these problems to a great extent.

In recent years, there has been significant maturation of biological waste air treatment research. This has resulted in a large number of studies concerning the performance and operation of the biofilters. Biofilter technology has a high potential for exhaust gas clean up, but as with many biological processes, the design requirements have not been fully appreciated. Interestingly, the fundamental processes involved during the elimination of a pollutant in a gas phase bioreactor are still very poorly understood.

The development of biofiltration has relied on the extensive experience gained in Europe, which has provided a substantial theoretical and practical knowledge base (Adler 2001). Research groups in several parts of the world, particularly in the Netherlands, Japan and the USA, are now developing novel applications for biofiltration. This expansion of applications is mainly due to several reasons: basic microbiological and biochemical research into the mechanisms of microbial degradation and the characterization of *microorganisms* suitable for achieving biofiltration; advances in filter bed media and packing design and bed loading techniques; development of models to predict biofilter behaviour during exposure to mixtures of volatile organic compounds, which may reduce the requirement for extensive pilot and field trials; development of alternative vapour-phase biological treatment systems, such as bioscrubbers and biotrickling filters and a growing understanding of the potential economic and environmental advantages of biofiltration within industry and the regulatory community (Adler 2001).

Biofilter technology was utilized in the field much before there was a basic understanding of its fundamental principles. This has resulted in several cases of unsuccessful or sub-optimum operation of large-scale bioreactors. Today, with recent advances in the understanding of the fundamental principles underlying biofiltration, promise exists for optimal operating conditions and better reactor design. However, a number of fundamental questions remain unanswered or require further clarification. Studies are required on the quantification of biomass turnover, biodegradation kinetic relationships and factors affecting these relationships ecology of biofilter microflora, the determination of the availability and cycles of pollutant, oxygen and essential nutrients. The above factors have been found to affect significantly the performance and long-term stability of biofilters, and therefore require further investigation in quantitative term. The expanding use of modern tools of biotechnology should be able to make it easier. The largest problem to overcome will be the translation of recent and future basic advances into real process improvements for biofiltration technology to mature from the mysterious black box reactor to a well-engineered process based on solid science rather than on trial and error.

Biofiltration technology for removal of odourous compounds from exhaust gases of pulp and paper industry has a great potential. Not much information directly related to the pulp and paper industry is available but extensive information is available on the biofiltration of organic compounds similar to those found in the exhaust gases of pulp and paper industry. Further studies are required for obtaining a better understanding of the mass transfer, heat transfer and reaction processes taking place within the biofilter beds. Extensive long-term studies of full scale biofilter systems would also be important in improving our understanding of biofilters used to remove volatile organic compounds from off gases generated in the paper industry. Extended studies of transient behaviour of biofilters are also required to provide the basic empirical knowledge essential for plant design, scale up and performance evaluation under real conditions.

The future of biofiltration depends on the regulatory requirements placed on industry. However, there are specific trends which will impact the market of biofiltration technology, and these trends are: Increased regulatory concern about emission of nitrogen oxides, which are emitted from thermal treatment processes. Biofilters do not create any additional nitrogen oxides; Increased public complaints about odourous emissions from public owned wastewater treatment plants, manufacturing industries, solid waste treatment facilities, etc.; Implementation of pollution prevention methodologies which has resulted in greater use of biodegradable solvents, reduced concentration of air emissions and emphasis on achieving zero discharge processes; and increased concern about emission of air contaminants, worker exposure to organics, emphasis on environmentally friendly and low-cost treatment technologies. The application of biofiltration technology has increased rapidly during the latter part of the twentieth century and will continue to grow throughout the twenty-first century. Though recent studies vary, depending on the underlying assumptions, the US biofiltration market for 1996 was estimated to be about $10 million (Kosteltz et al. 1996; Yudelson 1996). Potential markets for bio-filtration include: treatment of odours; treatment of volatile organic compounds and hazardous air pollutants and treatment of petroleum hydrocarbons.

Odour treatment is a significant portion of the marketplace. Industries that pro-duce odourous emissions include wastewater treatment plants, composting and sludge treatment facilities, foundries, pulp and paper plants and tobacco products manufacturing plants. In recent years, communities have begun to encroach near the fence lines of wastewater treatment plants. Wastewater treatment plants are treating increased flows, thereby increasing odour loads at the plant. Further, since flows are being pumped from greater distances, the age of the wastewater and its septicity is increasing, resulting in greater amounts of reduced nitrogen and sulphur com-pounds. In addition, water conservation has resulted in decreasing water flow rates with increased strength, which results in greater odour production. Many wastewa-ter treatment plants have begun to implement odour control strategies, and biofiltra-tion will play a major role in many such cases. Biotrickling filter technology was shown to be effective in treating odourous emissions from the "Zimpro" sludge heat treatment process, which has been known for creating very high intensity odours (Govind and Melarkode 1998).

Biofiltration of volatile organic compounds and hazardous air pollutants is an important problem in the wood products, pulp and paper and surface coating op-erations. In the case of surface coating operations, exposure of workers to organic chemicals, such as styrene, is an important issue. While attempts are being made to develop low volatile organic compounds emitting solvent formulations, some worker exposure is inevitable, and the use of biofiltration systems on the shop floor can reduce concentrations of organics in the ambient air. A pilot-scale study was conducted to demonstrate biotrickling filter technology for treating ethanol emis-sions from bakeries (Govind et al. 1998). Petroleum hydrocarbons are released during refining, transfer operations, from storage tanks, etc. Most of these hydrocar-bons consist of aliphatic and aromatic compounds, which are easily biodegraded in biofilters. Leaking underground storage tanks pose another environmental hazard, where the hydrocarbon contaminant can be separated from the soil and/or ground-water table using air sparging, bioventing or vapor extraction. The volatile hydro-carbons are transferred into the air phase, wherein they can be effectively treated using biofiltration.

As knowledge on biofiltration increases, and more pilot-scale studies are con-ducted, the market for biofiltration is expected to increase in the future. Increasing number of industries are already beginning to realize the potential advantages of biofiltration.

Biofiltration will play a major role in the treatment of organic and inorganic emis-sions from a variety of industrial and waste water treatment processes. Compost bio-filters are better suited for treatment of odours and low concentration (<25 ppmv) contaminants. Biotrickling filters have significant advantages over compost biofil-ters and are capable of handling significantly higher contaminant concentrations (20–5,000 ppmv). The major issues in biotrickling filters is the design of the support media and handling of biomass growth. Support media design has a significant impact on biotrickling filter performance. The market for biofilters will increase in the next millennium, as new applications arise in the future.

References

Adler SF (2001) Biofiltration a primer. Chem Eng Prog 97(4):33–41

Brauer H (1986) Biological purification of waste gases. Int Chem Eng 26(3):387–395

Gerrard AM, Metris AV, Paca J (2000) Economic designs and operation of biofilters. Eng Economist 45 259–270

Govind R, Bishop DF (1996) Overview of air biofiltration—basic technology, economics and integration with other control technologies for effective treatment of air toxics. Emerging solutions VOC air toxics control, Proc. Spec. Conf. (Pittsburgh, PA), 324–350

Govind R, Melarkode R (1998) Pilot-scale test of the biotreatment of odors from Zimpro™ sludge conditioning process. Report submitted to sanitation district No. 1, Fort Wright, KY, by PRD TECH, Inc., Florence, KY

Govind R, Fang J, Melarkode R (1998) Biotrickling filter pilot study for ethanol emissions control, a report prepared for the food manufacturing coalition for innovation and technology transfer, by PRD TECH, Inc., Florence, KY

Kosteltz AM, Finkelstein A, Sears G (1996) What are the 'real opportunities' in biological gas cleaning for North America? Proceedings of the 89th Annual Meeting & Exhibition of Air & Waste Management Association, A & WMA, Pittsburgh, PA, 96-RA87B.02

Leson G, Winer AM (1991) Biofiltration: an innovative air pollution control technology for VOC emissions. J Air Waste Manage 41:1045–1054

Naples BK (2010) Performance validation of the first North American shell-based biological air treatment system, odors and air pollutants 2010, Water Environment Federation

Ottengraph SPP, Van Denoever AHC (1983) Kinetics of organic compound removal from waste gases with a biological filter. Biotechnol Bioeng 25:3089–3102

Shareefdeen Z, Herner B, Singh A (2005) Biotechnology for air pollution control—an overview. In: Shareefdeen Z, Singh A (eds) Biotechnology for odor and air pollution control. Springer, Berlin

Soccol CR, Woiciechowski AL, Vandenberghel LPS, Soares M, Neto GN, Thomaz Soccol V (2003) Biofiltration: an emerging technology. Indian J Biotechnol 2(3):396–410

Thorsvold BR (2011) Biological odor control systems: a review of current and emerging technologies and their applicability nfo.ncsafewater.org/Shared Documents/Web Site Documents/Annual Conference/AC 2011 Papers/WW_T.pm_03.45_Thorsvold.pdf. Accessed 20 March 2014

Van Groenestijn JW, Hesselink Paul GM (1993) Biotechniques for air pollution control. Biodegradation 4:283–301

Van Lith C, Leson G, Michelson R (1997) Evaluating design options for biofilters. J Air Waste Manage Assn 47:37–48

Yudelson JM (1996) The future of the U.S. biofiltration industry. Proceedings of the 1996 Conference on Biofiltration (an Air Pollution Control Technology), Reynolds, F.E., Ed., The Reynolds Group, Tustin, CA, p 1

Index

P. Bajpai, *Biological Odour Treatment*, SpringerBriefs in Environmental Science,
DOI 10.1007/978-3-319-07539-6, © The Author(s) 2014